Distribution Category UC-98

Geothermal Energy in the Western United States and Hawaii: Resources and Projected Electricity Generation Supplies

September 1991

Energy Information Administration
Office of Coal, Nuclear, Electric and Alternate Fuels
U.S. Department of Energy
Washington, DC 20585

This report was prepared by the Energy Information Administration, the independent statistical and analytical agency within the Department of Energy. The information contained herein should not be construed as advocating or reflecting any policy position of the Department of Energy or of any other organization.

Contacts

Questions regarding the contents of this report may be directed to:

Nuclear and Alternate Fuels Division
Energy Information Administration, EI-53
U.S. Department of Energy
1000 Independence Avenue, S.W.
Washington, DC 20585

Questions of a general nature should be directed to Howard L. Walton (202/254-5500), Director of the Nuclear and Alternate Fuels Division; Dr. Z.D. (Dan) Nikodem (202/254-5550), Chief, Data Systems Branch; or Jim Disbrow (202/254-5558), Project Manager for this report.

Contributions were provided by the following employees of the Nuclear and Alternate Fuels Division:

Energy Assessment of Geothermal Resources:
Ron Fontanna
William Szymanski

Hydrothermal Electric Power Generation Technology:
Ron Fontanna

Current Status of the Industry:
William Szymanski

EIA Coordinator of the Sandia National Laboratory's study, *Supply of Geothermal Power from Hydrothermal Sources* (Appendix B):
John Carlin

Additional contributions to the report were provided by the following individuals:

Geothermal Geological Processes:
Dr. P.M. Wright
University of Utah Research Institute

Supply of Geothermal Power:
Susan Petty
Susan Petty Consulting, Inc.

Expert Reviewers:
Dr. D. Entingh
Meridian Corporation

Dr. Allan Jelacic
U.S. Department of Energy

Preface

At the request of the President of the United States, the Department of Energy is undertaking a comprehensive assessment of the Nation's energy future. The central themes of this assessment are "achieving balance among our increasing need for energy at reasonable prices, our commitment to a safer, healthier environment, our determination to maintain an economy second to none, and our goal to reduce dependence by ourselves and our friends and allies on potentially unreliable energy suppliers."[1] These concerns have been heightened as a result of the war in the Middle East and the international interest in global warming. The President has directed that a keystone of this strategy be continuing the successful policy of market reliance. Markets should be allowed to allocate scarce energy resources in the most efficient manner. In specific instances where markets cannot or do not work efficiently, the strategies would identify those barriers and recommend specific actions instrumental in alleviating them.

This project was conducted by the Energy Information Administration (EIA) to support the development of the National Energy Strategy and for preparation of the *Annual Energy Outlook*. The Office of Conservation and Renewable Energy of the Department of Energy, University of Utah Research Institute, Sandia National Laboratory, the Meridian Corporation, Petty Consulting, and Decision Analysis Corporation of Virginia all contributed to the data and analyses presented in this report.

EIA considers this report an initial step in expanding its data and analysis capabilities in the area of renewable energy. The report is intended for use by energy analysts, policymakers, Congress, State public utility commissions, and the general public.

[1]U.S. Department of Energy, *National Energy Strategy, Powerful Ideas for America*, First Edition 1991/1992, DOE/S-0082P (Washington, DC, February 1991).

Contents

	Page
Executive Summary	vii
1. Introduction	1
Background	1
Organization of the Report	1
2. Current Status of the Industry	7
Introduction	7
Electric Power Generation Technology	7
Electric Power Supplies	13
Market Development Trends	17
3. Electric Power Generation Potential	21
Introduction	21
Interlaboratory Task Force on Renewable Energy Assessment	21
Renewable Energy Technologies Subgroup Assessment	22
Sandia National Laboratory Assessment	23
A Comparison of Projections	25
Considerations Affecting Future Utilization of Geothermal Resources	25
Appendices	
A. Geothermal Resources	37
B. Sandia National Laboratory Study: Supply of Geothermal Power from Hydrothermal Sources	51
C. Descriptions of Liquid-Dominated Geothermal Power Plants	59
D. Geothermal Project Developers and Owners	61
Glossary	65

Tables

Page

ES1. Comparison of Estimates of U.S. Hydrothermal Electricity Resources and Potential Market Penetration viii
1. Significant Events in the Development of Geothermal Energy in the United States 3
2. Geothermal Generating Plants in the United States, December 31, 1990 8
3. Generation Costs of a 50-Megawatt Hydrothermal Plant in California 15
4. Hydrothermal Power Plants: Utility Generating Capacity Planned Additions and Announced Projects, 1991-1995 20
5. Interlaboratory Projections of Geothermal Electric Capacity and Generation, 1988-2030 22
6. Projections for Geothermal Capacity and Generation, 1990-2030 23
7. Comparison of Estimates of Potentially Installable Capacity from U.S. Hydrothermal Electricity Resources 24
8. Electricity Consumption Forecasts for the Western United States 31
A1. Classification of Geothermal Resources 41
A2. Estimated Physical Characteristics for Selected Identified High-Temperature Hydrothermal Resources . . 43
A3. Estimates of Hydrothermal Resources 49
B1. Projected Capacities of Hydrothermal Resource Sites in the Western United States and Hawaii 52
B2. Financial Assumptions of the Sandia Study 55
B3. Technology Improvement Assumptions 57

Figures

1. Wells Producing Geothermal Energy in California, 1980-1989 10
2. Schematic of a Geothermal Electricity Generating System for Vapor-Dominated Hydrothermal Resources 11
3. Schematic of a Single-Flash Geothermal Electricity Generating System for Liquid-Dominated Hydrothermal Resources 11
4. Schematic of a Double-Flash Geothermal Electricity Generating System for Liquid-Dominated Hydrothermal Resources 12
5. Schematic of a Binary Geothermal Electricity Generating System for Liquid-Dominated Hydrothermal Resources 12
6. Annual U.S. Hydrothermal Electric Capacity 13
7. Exploratory Geothermal Energy Wells in California, 1976-1989 18
A1. Schematic of the Earth's Interior 38
A2. Schematic of Tectonic Plate Movements 38
A3. Tectonic Regions in North America and the Eastern Pacific Ocean 40
A4. Cross-section of the Earth Showing Source of Geothermal Energy 42
A5. Hot Dry Rock (HDR) Geothermal System Concept for Low-Permeable Formations 44
A6. Selected Geothermal Resource Areas in the Western United States 45
A7. Heat Flow Contours for the Western United States 46
A8. Stages in the Discovery and Development of Geothermal Resources 47
B1. Supply of Geothermal Power: Current Technology, Identified Resources 56
B2. Potential Supply of Geothermal Power: Current Technology, Identified and Estimated Unidentified Resources 56
B3. Potential Supply of Geothermal Power: Technology Improvement Assumptions, Identified Resources . . 58
B4. Potential Supply of Geothermal Power: Technology Improvement Assumptions, Identified and Estimated Unidentified Resources 58

Executive Summary

Geothermal energy comes from the internal heat of the Earth, and has been continuously exploited for the production of electricity in the United States since 1960. Currently, geothermal power is one of the ready-to-use baseload electricity generating technologies that is competing in the western United States with fossil fuel, nuclear and hydroelectric generation technologies to provide utilities and their customers with a reliable and economic source of electric power. Furthermore, the development of domestic geothermal resources, as an alternative to fossil fuel combustion technologies, has a number of associated environmental benefits.

This report serves two functions. First, it provides a description of geothermal technology and a progress report on the commercial status of geothermal electric power generation. Second, it addresses the question of how much electricity might be competitively produced from the geothermal resource base.

Current Status of the Geothermal Industry

Geothermal resources can be subdivided into four categories: (1) hydrothermal,[2] (2) geopressured-geothermal, (3) hot dry rock, and (4) magma. All existing commercial electric power generation comes from hydrothermal resources. Electricity generation utilizing dry steam (i.e., vapor-dominated) is the oldest and simplest geothermal technology, involving the passage of the steam directly from a geothermal reservoir to a turbine. This technology has been successfully used to exploit these relatively rare resources. Recently, several electricity generation technologies have been developed to efficiently extract heat from the far more abundant liquid-dominated resources. These include the single-flash, double-flash, and binary-cycle systems (discussed in Chapter 2).

The U.S. geothermal power industry has been in existence for over three decades, and by the end of 1990 there were 2,719 megawatts of electric capacity installed at 70 hydrothermal plants located in California, Nevada, and Utah. Between 1984 and 1990, installed hydrothermal capacity increased by over 80 percent. The most economic and heavily developed geothermal site is a steam reservoir, The Geysers, in California. At year-end 1990, installed capacity at The Geysers was 1,866 megawatts. The remaining 30 percent of the Nation's installed geothermal capacity taps liquid-dominated reservoirs at other sites.

Currently, power contracts are in place to bring another 278 megawatts of geothermal capacity into production by the end of 1995, principally in California and Nevada. An additional 386 megawatts of geothermal capacity are in the planning stages.

Geothermal Potential

Expanded use of hydrothermal resources represents a significant portion of the geothermal potential in the 1990 through 2030 period. However, the use of the other geothermal resource types, particularly hot dry rock, could become significant toward the end of the period.

In 1979, the United States Geological Survey (USGS) estimated the potential electrical generation capacity from hydrothermal energy. Based on what was known about the identified hydrothermal resources, the USGS estimated that 23,000 megawatts of annual electrical generation capacity could potentially be supported over a 30-year lifespan of the resources. The USGS estimated that between 72,000 and 127,000 megawatts of additional electrical generation capacity could be supported by undiscovered resources (Table ES1). Hydrothermal resource assessments are typically reported as the total number of annual megawatts of electrical power that is potentially producible over a period of time. These estimates were made with no consideration for the costs to extract heat from the resource base and produce commercially competitive power. Additionally, the USGS estimates are limited by having only considered resources above 150°C.

In a 1990 study, the Interlaboratory Task Force on Renewable Energy developed projections—based on consensus judgements of expected technology costs and

[2]Definitions for this and other technical terms can be found in Appendix A and in the Glossary.

Table ES1. Comparison of Estimates of U.S. Hydrothermal Electricity Resources and Potential Market Penetration
(Megawatts)

	Installed and Potentially Installable Capacity			Resource Base
	1990	2010	2030	
Installed	2,719			
Interlaboratory Task Force Projections[a]		5,900	10,600	
Renewable Energy Subgroup Assessment[b]		10,650	23,400	
Sandia Resource Assessment[c]				
Current Technology, Identified Resources (Base Case)[d]		6,000	14,000	
Improved Technology, Identified and Estimated Unidentified Resources[e]		15,500	44,000	
USGS Resource Assessment[f,g]				
Identified	NA[h]	NA[h]	NA[h]	23,000[i]
Identified Plus Undiscovered	NA[h]	NA[h]	NA[h]	95,000-150,000[i]

Sources: Compiled by the Office of Coal, Nuclear, Electric and Alternate Fuels:
[a]U.S. Department of Energy, *The Potential of Renewable Energy*, An Interlaboratory White Paper, SERI/TP-260-3674 (Solar Energy Research Institute, Golden, CO, March 1990), Business As Usual Case, Table C-1, p. C-7. The estimates for 2010 and 2030 include 670 megawatts and 4,710 megawatts, respectively, of non-hydrothermal resources (i.e., hot dry rock, geopressurized, and magma).
[b]Energy Information Administration, *Renewable Energy Excursion: Supporting Analysis for the National Energy Strategy*, SR/NES/90-04 (Washington, DC, December 1990), calculated from the Baseline Case, Table 2, p. 22. In 2010, the 10,650 megawatts were estimated to be competitive with other energy sources at less than 6 cents per kilowatthour for delivered electricity.
[c]Petty, S., Livesay, B.J., and Geyer, J., *Supply of Geothermal Power from Hydrothermal Sources: A Study of the Cost of Power Over Time*, prepared for the U.S. Department of Energy (Sandia National Laboratory, 1991) (Draft).
[d]Capacity was estimated to be available at 6 cents per kilowatthour.
[e]Capacity was estimated to be available at 12.5 cents per kilowatthour.
[f]Economic and market factors not considered.
[g]Muffler, L.J.P., editor, *Assessment of Geothermal Resources of the United States—1978*, U.S. Geological Survey Circular 790 (1978), Table 4, p. 41. Excludes reservoirs in Cascades and National Parks.
[h]Not applicable. The estimates of recoverable energy are unbounded by time.
[i]Estimates of recoverable energy unbounded by time. Data represent annual electrical generation capacity potentially obtainable over a 30-year lifespan of the resources.

penetration rates—reaching 10,600 megawatts of geothermal electric power capacity by the year 2030, quadrupling the 2,719 megawatts in operation at the end of 1990. In a study later in 1990, conducted by the Renewable Energy Subgroup of the National Energy Strategy Modeling Group, geothermal electric power capacity was projected to reach 23,400 megawatts by the year 2030. These results were based on a methodology that compared alternative baseload technologies using life-cycle costs to estimate market shares of various electricity generating technologies.

In order to improve the capability to project geothermal electric power capacity, the Energy Information Administration in conjunction with the Office of Conservation and Renewable Energy, DOE, sponsored a study by the Sandia National Laboratory to extend the USGS analysis to include moderate temperature hydrothermal resources and include new information made available after 1979. A major goal of this project was to estimate the costs to produce various levels of geothermal electric capacity, so-called "supply curves." Supply curves were developed based on geothermal reservoir characteristics, potential technology development, and expectations concerning operating performance. Under the improved technology assumptions, utilizing both the identified and estimated unidentified resources and maximum busbar costs[3] of

[3]The power plant "bus" or "busbar" is that point beyond the generator but prior to the voltage transformation point in the plant switchyard. The *busbar cost* represents the cost per kilowatthour to produce electricity, including the cost of capital, debt service, operation and maintenance, and fuel.

12.5 cents per kilowatthour, the Sandia study found that up to 44,000 megawatts of geothermal electric capacity could be developed by 2030 (Table ES1). As a conservative estimate with no advancements in technology, utilizing only the identified resources and assuming costs that are competitive with current alternative baseload technologies, Sandia found that 14,000 megawatts of geothermal capacity could be available at 6 cents per kilowatthour by 2030.

The Sandia study illustrates the uncertainties inherent in projecting the future use of geothermal energy for generating electricity. The prospects for increased exploitation of hydrothermal resources are dictated by conditions applicable to any energy technology: access to secure, long-term fuel supplies with acceptable environmental impacts and predictable costs; the maintenance of capital and operating costs at a level that produces competitively priced energy relative to other energy technologies; and the extent of political and institutional barriers that add uncertainty to the business climate.

In the case of geothermal technology, prospects for development include: (1) finding the large, but undiscovered, resource base; (2) accurately predicting long-term operational performance; (3) responding to competitive pressures from other energy technologies; (4) meeting environmental and other constraints on facility siting, water supply, waste effluent disposal, and power transmission; (5) continued improvement of the technology through experience and research; and (6) satisfying market demand. Uncertainties associated with these conditions lead to as much as a fivefold difference in estimates of the future use of geothermal energy for generating electricity.

1. Introduction

Background

Geothermal energy is the naturally occurring heat from the interior of the earth. Volcanoes are the most spectacular manifestation of the earth's capacity to provide heat. Other, less dramatic physical evidence is embodied in geysers, fumaroles,[4] and hot springs. The earliest use of geothermal energy by man was for bathing, which has been a cultural phenomenon for millennia. Thermal water has also been used for aquaculture,[5] greenhousing,[6] industrial process heat, such as an onion dehydration plant in Brady Hot Springs, Nevada,[7] and for space heating such as the district heating system in Boise, Idaho, developed in the early 1900's and expanded in the 1980's.[8] Electricity was first produced from geothermal resources at Larderello, Italy, in 1913.[9] In the United States, electricity was first produced at The Geysers, near Clear Lake in northern California, in 1960.[10]

In the United States, public sector involvement in the geothermal industry began with the passage of the Geothermal Steam Act of 1970 (Public Law 91-581). This Act authorized the Department of Interior to lease geothermal resources on Federal lands. The industry was subsequently influenced by other events, including the Organization of Petroleum Exporting Countries (OPEC) embargo of 1973 and the passage of the Federal Geothermal Energy Research, Development and Demonstration Act of 1974 (Public Law 93-410). This Act established a Federal interagency task force—the Geothermal Energy Coordination and Management Project—providing for research, development, and demonstration of geothermal energy technologies, and establishing a loan guaranty program for financing geothermal energy development.

A chronology of significant events in the development of geothermal energy in the United States is presented in Table 1. The long relationship of joint activities supported by both the geothermal industry and the Federal Government is evidenced in this table. This chronology can be examined in conjunction with information presented in Chapter 2 (Table 2) showing the growth in the number of operational, geothermal power plants. Interest in geothermal resources was further increased in response to market creation opportunities brought about as a result of the passage of the Public Utility Regulatory Policies Act of 1978 (PURPA) (Public Law 95-617). This Act authorized the Federal Energy Regulatory Commission to require utilities to offer to buy electrical power from qualifying facilities at the utility's full avoided cost.

Organization of the Report

This report provides information on the status of the geothermal energy industry, the electric power generation potential from geothermal resources, and a description and quantification of the geothermal resource base.

Chapter 2 describes the current status of the industry. Various types of technologies for generating electricity from hydrothermal resources are discussed, as are production statistics and associated power generation capacities, geothermal power marketing, the

[4]A *fumarole* is a vent from which steam or gases issue—a geyser or spring that emits gases.

[5]Johnson, W.C., *Culture of Freshwater Prawns Using Geothermal Waste Water* (Geo-Heat Center, Oregon Institute of Technology, Klamath Falls, OR, 1978).

[6]Rafferty, K., *Some Considerations for the Heating of Greenhouses with Geothermal Energy* (Geo-Heat Center, Oregon Institute of Technology, Klamath Falls, OR, 1985).

[7]Austin, J.C., CH2M Hill, Inc., *Direct Utilization of Geothermal Energy Resources in Food Processing*, Final Report, May 17, 1978 - May 31, 1982, Report No. DOE/ET/28424-6, Cooperative Agreement No. DE-FC07-78ET28424 (May 1982).

[8]Hanson, P.J., Boise Geothermal, "Boise Geothermal District Heating System Final Report March 1979-September 1985," Report No. DOE/ET/27053-6, Cooperative Agreement No. DE-FC07-79ET27053 (October 1985).

[9]Armstead, H.C.H., *Geothermal Energy: Its Past, Present, and Future Contribution to the Energy Needs of Man*, 2nd edition (London: E.F. Spoon, 1983), p. 5.

[10]Northwest Power Planning Council, *Assessment of Geothermal Resources for Electric Generation in the Pacific Northwest* (Portland, OR, 1989), p. 3.

environmental impact of geothermal power, and institutional and other constraints on the industry. Chapter 3 reviews the assessments of electricity generation contribution from geothermal resources conducted by the Interlaboratory Task Force on Renewable Energy and the National Energy Strategy Modeling Group's Renewable Energy Technology Subgroup. In addition, the potential electricity supplies that could be extracted from hydrothermal resources, irrespective of demand, were estimated by Sandia National Laboratory. The principal market and technological factors that are expected to influence the development and production of electricity from geothermal resources are also briefly discussed.

Appendix A gives an overview of the physical characteristics and the potential energy content of geothermal resources. Included as Appendix B is a detailed description of the Sandia Laboratory Geothermal Power Study. Brief descriptions of selected operational liquid-dominated geothermal power plants are given in Appendix C. A list of geothermal project developers and owners is presented in Appendix D. Technical terms are defined in the Glossary.

Geothermal Power Producers Win Environmental Awards

The California Energy Co., Ormat Energy Systems, Inc., and Pacific Gas & Electric Co., have received prestigious awards related to their geothermal operations. At a White House ceremony in April 1990 hosted by President Bush, California Energy's Coso Project received the National Environmental Award for its role in reducing greenhouse gases and ozone depleting chemicals. The award, sponsored by Renew America, a coalition of environmental advocacy groups, was judged from more than 1000 nominees by the officers of many leading U.S. environmental groups. The reductions in pollutants recognized by the award are achieved by a modification to the flash steam power process in which all noncondensible gases present in the geothermal fluid are injected back to the subsurface, reducing surface emission to virtually zero.

The Ormat award, presented by the American Society of Mechanical Engineers, recognized the company's proprietary development of technology that economically generates electric power using lower temperature heat sources that is viable today with other geothermal technologies. This ability may open up large additional quantities of geothermal resources in this country and abroad for power development. The closed binary systems generate no airborne emissions, and when air cooled condensers are used, there is no consumption of surface or ground water. ASME's Energy Resources Technology Awards recognize technologies less than five years old that serve to enhance the industrialization of the energy resources industry and that contribute to the improvement of the U.S. position in the world market.

The Pacific Gas & Electric Co. received California's first Air Pollution Reduction Award for developing and using a process to remove hydrogen sulfide gas from geothermal steam and reducing hazardous waste by as much as 90 percent at The Geysers power plant in Sonoma and Lake Counties, California. The process significantly reduces the volume of chemicals used at the plant.

Source: Excerpted from U.S. Department of Energy, *Geothermal Progress Monitor*, No. 12 (December 1990), pp. 15-16.

Table 1. Significant Events in the Development of Geothermal Energy in the United States

Date	Event
1891	District heating implemented in Boise, Idaho
1900	Hot water provided to homes in Klamath Falls, Oregon
1916	Steam power harnessed for electricity generation at The Geysers resort, California
1927	First exploratory geothermal wells drilled in the Imperial Valley, California by Pioneer Development Company
1959	Small pilot plant operated near Niland, California, on Sinclair No. 1 well
1960	First commercial electricity generated continuously from dry steam at The Geysers, California
1970	Geothermal Steam Act (Public Law 91-58) passed
1973	National Science Foundation (NSF) became lead agency for Federal Geothermal Programs
	U.S. Geological Survey (USGS) and NSF prepared the first Federal Geothermal Programs Plan
1974	Geothermal Energy Research Development and Demonstration (RD&D) Act (Public Law 93-410) passed which included the establishment of the Geothermal Loan Guaranty Program (GLGP)
1975	Energy Research and Development Administration (ERDA) formed; Division of Geothermal Energy (DGE) formed primarily from NSF staff to manage an RD&D program
	USGS released first national geothermal resource estimate and inventory
	Installed capacity reaches 500 megawatts (MW), all at The Geysers, California
1977	Department of Energy (DOE) formed; DGE continued to manage the RD&D program
1978	Energy Tax Act (Public Law 95-618) passed providing energy tax credits
	Public Utility Regulatory Policies Act (Public Law 95-617) (PURPA) enacted. Environmental Protection Agency (EPA) issued pollution control guidelines for geothermal energy development
	Hot dry rock reservoir created and tested in New Mexico
	First geothermal crop-drying plant built in Nevada
1979	USGS released updated national geothermal resource estimates and inventory
	Interdevelopmental task force recommended measures to speed federal leasing
	U.S. Navy awarded a contract to develop 75 megawatts at the Coso Hot Springs known geothermal resource area on the Naval Weapons Center, China Lake, California

See source note at end of table.

Table 1. Significant Events in the Development of Geothermal Energy in the United States (Continued)

Date	Event
1980	Federal Energy Regulatory Commission (FERC) issued regulations (18 CFR 292) establishing hydrothermal geothermal resources as renewable resources and geothermal facilities as qualifying facilities
	World's largest single geothermal power unit (132 megawatts) generated electricity at The Geysers, California
	10-megawatt flash-steam plant built by industry at Brawley, California
	First electric power from hot dry rock produced at Fenton Hill, New Mexico
	First geothermal ethanol plant began production at La Grande, Oregon, under private funding
	First five DOE-sponsored field demonstrations of direct heat applications became operational
	First deep geothermal reservoir confirmation well drilled in Atlantic Coastal Plain near Crisfield, Maryland
	Crude Oil Windfall Profits Tax Act (Public Law 96-223) passed, providing tax credit increase for geothermal equipment
	Energy Security Act (Public Law 96-294), containing Title VI, "The Geothermal Energy Act of 1979," was passed
1981	First U.S. geothermal electric generation plant outside the 48 contiguous States brought on-line in the Puna resources area in Hawaii
	The Insurance Company of North America began offering insurance against the financial risk of reservoir failure
	The practical demonstration of generating electricity from moderate-temperature geothermal fluids was accomplished at Raft River, Idaho
	A mobile well-head generator with a net output of 1.6 megawatts was installed at Roosevelt Hot Springs, Utah
	USGS research drilling at Newberry Volcano, Oregon indicated for the first time that temperatures (265°C at 3,057 ft) sufficient for electrical production existed in the Cascade Mountains
	FERC issued amendments to its regulations for qualifying small power production facilities incorporating the provisions of the Energy Security Act relating to nonutility geothermal facilities
1982	A 10-megawatt flash plant utilizing hypersaline brine began operation at the Salton Sea KGRA, California
	An 80-megawatt geothermal electric power plant to be constructed by Occidental Geothermal, Inc., in Lake County, California, and a 49-megawatt geothermal electric power plant to be constructed by Republic Geothermal, Inc., and the Parsons Corporation in the Imperial Valley, California were certified by the FERC as qualifying facilities. Magma Power Company and Magma Development Corporation issued a public notice of self-qualifications for an existing 11-megawatt geothermal power plant located in East Mesa, California
	In an effort by the Department of Interior to accelerate the geothermal leasing program, a record 16 competitive lease sales were held in which 578,656 acres were offered
	USGS completed the first quantitative national assessment of low-temperature (<90°C) geothermal resources of the United States
	Total U.S. installed geothermal capacity reaches 1,000 megawatts

See source note at end of table.

Table 1. Significant Events in the Development of Geothermal Energy in the United States (Continued)

Date	Event
1983	Federal leasing regulations were rewritten, resulting in the deletion of burdensome, counterproductive requirements
	The largest massive hydraulic fracture operation in North America created a second hot dry rock reservoir at the Fenton Hill, New Mexico, site
1984	The first commercial electric power from federal lands outside California was generated, with 20 megawatts brought on-line by Philips Geothermal in the Roosevelt Hot Springs KGRA in Utah
1985	Lease acreage limit increased by DOI from 20,480 to 51,200 acres per State
	Total U.S. installed geothermal capacity reaches 2,000 megawatts
1986	The Salton Sea scientific well was drilled and cored to 10,564 ft; cuttings, fluid samples and geophysical well logs were obtained, and preliminary flow tests conducted
	Congress acted to preclude geothermal leasing where, in the judgment of the Secretary of the Interior, development would result in significant adverse effects to significant thermal features in national parks
	The 10-megawatt flash-steam plant at Brawley, California, ceased operation
1987	DOE cost-shared with industry the drilling of three deep thermal gradient test wells within the Cascades volcanic area of the Pacific Northwest--which demonstrated the utility of this technique in identifying underlying hydrothermal features such as those beneath the Cascades
	Navy Geothermal Plant Number One, Unit Number One, at Naval Weapons Center (NWC), China Lake California, began delivering power to the public utility grid. Unit One is rated at 25-megawatts capacity. NWC's peak power demand is 20 megawatts. Negotiations were underway for an additional 135 megawatts at planned Units Two through Six
	The decline in production of steam at The Geysers begins
1988	Geothermal Steam Act Amendments of 1988 were signed into law (Public Law 100-443). This law significantly modifies the geothermal leasing programs by providing two 5-year extensions of the primary term if a geothermal steam has not been produced in commercial quantities by the end of the primary term. The Act further provides protection for significant thermal features in units of the National Park System from the effects of geothermal development
	Navy Geothermal Plant Number One, Units Number Two and Three, Naval Weapons Center, China Lake, California was completed with the three generating units capable of generating 80 megawatts. Construction started on a second 80-megawatt plant with production scheduled for early 1990
1989	Eight geothermal electric plants became operational with a combined total of 295 megawatts, including Unocal's Salton Sea Unit Three plant which utilizes the crystallizer clarifier process developed initially by the Department of Energy
	Operation and testing of a joint DOE/Electric Power Research Institute (EPRI) heat/methane hybrid power system (HPS) using geopressured brine was begun at the Pleasant Bayou, Texas, well site. Electricity from the 0.98-megawatt unit is sold to a local utility
	The first exploratory well to be sited directly over a suspected magma body was spudded and drilled to a depth of 2,568 feet by DOE. A corehole was then drilled as part of the Continental Scientific Drilling Program

See source note at end of table.

Table 1. Significant Events in the Development of Geothermal Energy in the United States (Continued)

Date	Event
	The binary demonstration plant at Heber, California, ceased operation
	The demonstration plant on the Puna Coast of Hawaii ceased operation
1990	Three geothermal power producers win environmental awards (see box above)

Source: Compiled by the Office of Coal, Nuclear, Electric and Alternate Fuels. Modified from Budget and Planning Working Group of the Interagency Geothermal Coordinating Council, *Fourteenth Annual Interagency Geothermal Coordinating Council Report for Fiscal Year 1989* (April 24, 1990), pp. 2-4.

2. Current Status of the Industry

Introduction

Geothermal resources suitable for electric power generation come in four different forms: hydrothermal, geopressured, hot dry rock, and magma. A summary description of these resources from a geological perspective can be found in Appendix A of this report. To date the commercial production of electricity has come only from hydrothermal resources, mainly at The Geysers, a vapor-dominated (steam) hydrothermal resource located some 70 miles north of San Francisco, California. Approximately 70 other Known Geothermal Resource Areas (KGRAs) have liquid-dominated hydrothermal resources. Several of these have been developed commercially over the past 15 years.

Production statistics and associated power generation capacities for each hydrothermal power plant are shown in Table 2. The installed net capacities reflect the capability of each plant's equipment to produce electricity. The electricity production figures are the best available data for actual production in 1989. The date of initial operation for each plant indicates when the plant was connected to the grid. While the costs associated with bringing a plant into operation vary considerably from one location to another, an estimate of costs representative of two future plant types is given in Table 3. Table 4 describes the expansion of the industry, detailing the somewhat speculative plans of State and industry contacts, as well as the known contracts for additional power.

Electric Power Generation Technology

Currently, all commercial geothermal electric power generation comes from hydrothermal resources (Table 2). The only vapor-dominated hydrothermal resource, The Geysers, has one field shared by ten companies. These companies utilize technologies associated with extracting energy from the steam. All other commercial generation comes from liquid-dominated resources which require different specialized technologies. While all of these commercially productive known geothermal resource areas tend to be located in fairly remote areas, they are still close to existing transmission networks. Since hot water and steam can be transported only a few thousand yards without significant heat loss, the most efficient exploitation of high-grade hydrothermal resources is achieved by on-site conversion of thermal energy to electricity.

Technology for Vapor-Dominated Resources

Electricity generation utilizing dry steam is an old and simple geothermal technology. In 1990, 1,866 megawatts of the 2,719 megawatts (approximately 69 percent) of installed geothermal generating capacity in the United States (Table 2) was derived from The Geysers. This unique natural resource has been the primary site of the domestic geothermal industry since it was first tapped in 1960. The majority of producing wells have historically been located at The Geysers (Figure 1). This trend is changing as development occurs because vapor-dominated resources are geologically limited. At The Geysers dry steam passes directly from the reservoir to a turbine (Figure 2). The steam is routed through a condenser and the condensate is either used for cooling tower charge makeup or is injected into the underground reservoir.[11] Hydrogen sulfide gas must be removed prior to the discharge of noncondensable gases into the atmosphere. Until 1979, commercial exploitation of geothermal resources was limited to dry steam technology. While commercial-grade vapor-dominated resources are rare, they remain the most economic source of geothermal energy for electricity generation.

Technology for Liquid-Dominated Resources

Liquid-dominated resources are far more widespread than vapor-dominated resources.[12] These resources are generally liquid in the ground, where they exist at or above the boiling point for the reservoir pressure. In order to extract energy from the resource, the fluid is produced from wells either through a high internal pressure drive or by pumping.

[11]Northwest Power Planning Council, *Assessment of Geothermal Resources for Electric Generation in the Pacific Northwest* (Portland, OR, 1989), p. 9.

[12]Williams, S., and Porter, K., *Power Plays: Geothermal* (Investor Responsibility Research Center, 1989), p. 165.

Table 2. Geothermal Generating Plants in the United States, December 31, 1990

State/Location/Plant Name[a]	Net Capacity (Megawatts)	1989 Generation[b] (Thousand MWh)	Date of Initial Operation	Qualifying Facility[c]
California				
Coso Hot Springs				
Navy 1, Unit 1	28	[d]629	7/87	Yes
Navy 1, Unit 2	28		11/88	Yes
Navy 1, Unit 3	28		11/88	Yes
Navy 2, Unit 4	28	[e]0	11/89	Yes
Navy 2, Unit 5	28		12/89	Yes
Navy 2, Unit 6	28		12/89	
BLM, East 1	28	[f]265	12/88	Yes
BLM, East 2	28		12/88	Yes
BLM, West	28	[g]0	8/89	Yes
East Mesa				
GEM 1 (formerly BC McCabe)	13	18	11/79	Yes
Ormesa 1	24	[h]234	12/86	Yes
Ormesa 1E	4		12/88	Yes
Ormesa 1H	6	[e]0	12/89	
Ormesa II	17	146	6/87	Yes
GEM 2	19	59	5/89	Yes
GEM 3	19	43	6/89	Yes
The Geysers/Sonoma County				
PG&E Units 2-14, 16-18, 20	1,291	8,053	3/63-10/85	
West Ford Flat	29	216	12/88	Yes
Bear Canyon Creek	22	165	10/88	Yes
Northern California Power Agency				
NCPA 1	110	664	1/83	
NCPA 2	110	644	10/85	
Santa Fe Geothermal (formerly Occidental)	88	690	4/84	
Sacramento Municipal Utility District				
SMUDGEO 1	72	589	12/83	
Coldwater Creek 1	62	[i]410	6/88	
Coldwater Creek 2	62		7/88	
Joseph Aidlin Plant	20	[g]0	5/89	Yes
Heber				
Heber Dual Flash	47	350	8/85	
Mono-Long Valley				
Mammoth Pacific 1	10	67	2/85	Yes
Mammoth Pacific 2	12	[e]0	12/90	Yes
PLES Unit 1	12	[e]0	12/90	Yes
Salton Sea				
Salton Sea 1	10	73	6/82	Yes
Salton Sea 2	18	[e]0	3/90	Yes
Salton Sea 3	51	333	2/89	Yes
Vulcan	32	288	12/85	Yes
Del Ranch	36	326	12/88	Yes
Elmore 1	36	291	12/88	Yes
Leathers 1	36	[e]0	12/89	Yes

See footnotes at end of table.

Table 2. Geothermal Generating Plants in the United States, December 31, 1990 (Continued)

State/Location/Plant Name[a]	Net Capacity (Megawatts)	1989 Generation[b] (Thousand MWh)	Date of Initial Operation	Qualifying Facility[c]
California (Continued)				
Wendell-Amedee				
Amedee Geothermal	2	11	11/88	Yes
Wineagle	1	[e]0	9/85	Yes
Honey Lake	30	[j]0	1/88	Yes
Subtotal	**2,553**			
Nevada				
Beowawe Hot Springs				
Beowawe	17	81	12/85	Yes
Brady Hazen				
Desert Peak	9	73	12/85	Yes
Dixie Valley				
Oxbow	57	425	2/88	Yes
San Emidio Desert				
Empire Geothermal Project	3	15	12/87	Yes
Stillwater/Soda Lake				
Soda Lake Geothermal Project	3	19	12/87	Yes
Stillwater Geothermal Project	11	[g]0	4/89	Yes
Soda Lake II	13	NA	12/90	
Steamboat Springs				
Steamboat Geothermal Project I	7	49	10/86	
Yankee/Caithness Joint Venture	13	84	2/88	Yes
Wabuska				
Wabuska	1	8	9/84	Yes
Subtotal	**134**			
Utah				
Cove Fort				
Cove Fort Geothermal 1	2	[g]0	9/85	Yes
Cove Fort Steam 1	2	[g]0	9/88	Yes
Cove Fort Steam 2	8	[g]0	11/89	
Roosevelt Hot Springs-Milford				
Blundell Unit 1	20	173	7/84	Yes
Subtotal	**32**			
U.S. Total for 70 Plants	**2,719**	**15,491**		

Notes: Locations are designated as Known Geothermal Resource Areas (KGRAs). Totals may not equal sum of components due to independent rounding.

Sources: Compiled by the Office of Coal, Nuclear, Electric and Alternate Fuels:

[a]*Geothermal Generating Plants in the United States* (March 1991).

[b]B. Swezy, Solar Energy Research Institute, personal correspondence to D. DeJarnette, 1991.

[c]Entingh, D., Meridian Corporation, April 9, 1991, personal correspondence. The Public Utility Regulatory Policies Act of 1978 (PURPA) required utilities to offer to buy electrical power from qualifying facilities.

[d]Includes all of Navy 1 (Units 1, 2, and 3).

[e]Units not operational in 1989.

[f]Includes BLM (East 1 and 2).

[g]Not available.

[h]Includes Ormesa 1 and 1E.

[i]Includes Coldwater Creek 1 and 2.

[j]Honey Lake Plant is a wood-waste cogeneration plant that uses geothermal fluids only to preheat boiler feedwater.

Figure 1. Wells Producing Geothermal Energy in California, 1980-1989

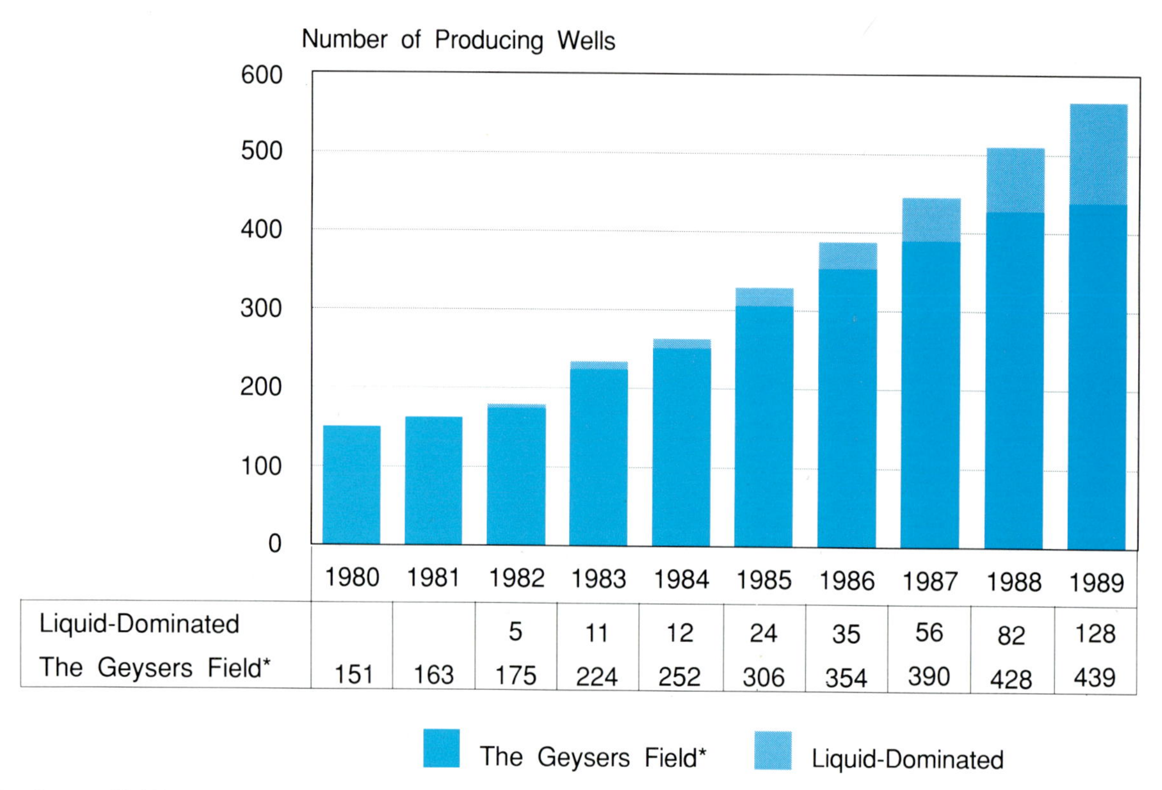

	1980	1981	1982	1983	1984	1985	1986	1987	1988	1989
Liquid-Dominated			5	11	12	24	35	56	82	128
The Geysers Field*	151	163	175	224	252	306	354	390	428	439

*The Geysers Field is a vapor-dominated geothermal field.

Source: California Department of Conservation, Division of Oil and Gas, *75th Annual Report of the State Oil and Gas Supervisor, 1989*, p. 154.

Several technologies have been developed to extract heat efficiently from liquid-dominated resources. The specific technology utilized at a given site depends on resource characteristics such as water temperature, noncondensible gas content, and salinity. Included among hot water technologies are single-flash systems (Figure 3), double-flash systems (Figure 4), and binary-cycle systems (Figure 5).[13,14,15]

- **Single-flash Systems.** In a single-flash system, fluid is allowed to boil at the surface in a one-stage production separation (Figure 3). Depending on the resource temperature, a fraction of the hot water "flashes" to steam when exposed to the lower pressure within the separator. The steam is then passed through a turbine to generate power. Typically, the liquid fraction is then injected back into the reservoir, but other beneficial uses, such as district heat, are possible.

- **Double-flash Systems.** Double-flash technology imposes a second-stage separator onto a single-flash system. The liquid remaining after the first-stage separation is flashed once more (Figure 4). This second-stage steam has a lower pressure and is either put into a later stage of a high-pressure turbine or through a second lower pressure turbine. Double-flash technology is in the range of 10 to 20 percent more efficient than single-flash technology.

- **Binary Cycle Systems.** Binary cycle technology incorporates two distinct closed fluid loops to generate electricity (Figure 5). The first loop passes hot water from the reservoir to a heat exchanger. A second loop moves a cold liquid phase working fluid (e.g., isobutane or some other hydrocarbon matched to the reservoir temperature) to the heat exchanger. Upon heating, the working fluid rapidly reaches its boiling point. The vaporized working

[13]Armstead, H.C.H., *Geothermal Energy: Its Past, Present, and Future Contribution to the Energy Needs of Man*, second edition (E.F. Spoon, London, 1983), pp. 168-170.

[14]Northwest Power Planning Council, *Assessment of Geothermal Resources for Electric Generation in the Pacific Northwest* (Portland, OR, 1989), pp. 9-11.

[15]Williams, S., and Porter, K., *Power Plays: Geothermal* (Investor Responsibility Research Center, 1989), pp. 174-175.

Figure 2. Schematic of a Geothermal Electricity Generating System for Vapor-Dominated Hydrothermal Resources

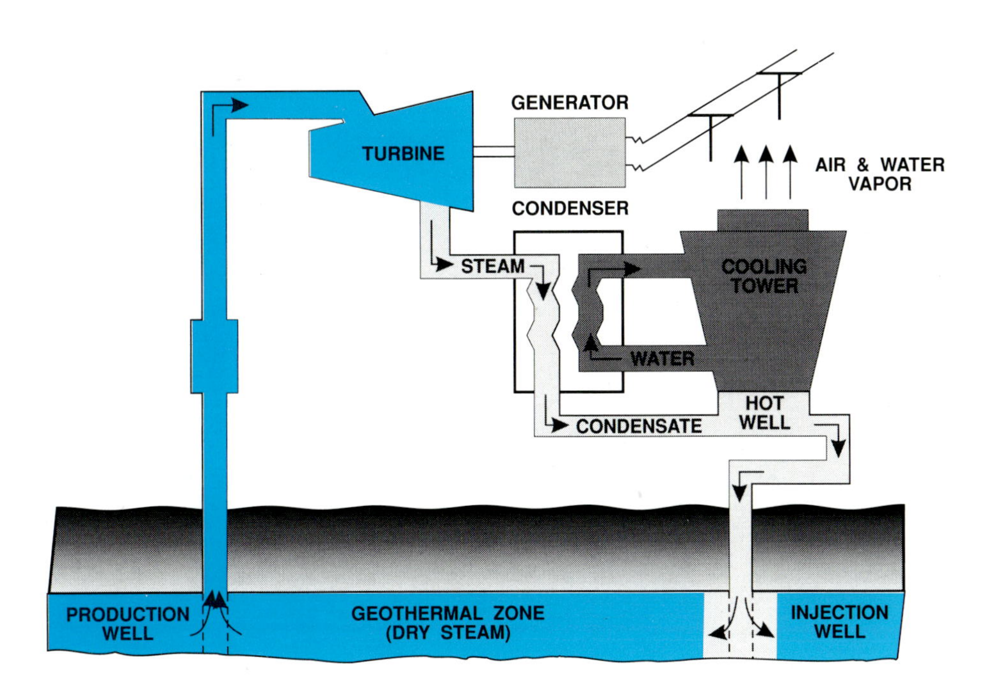

Source: Petroleum Information Corporation, *The Geothermal Resource* (A.C. Nielsen Co., 1979), p. 40.

Figure 3. Schematic of a Single-Flash Geothermal Electricity Generating System for Liquid-Dominated Hydrothermal Resources

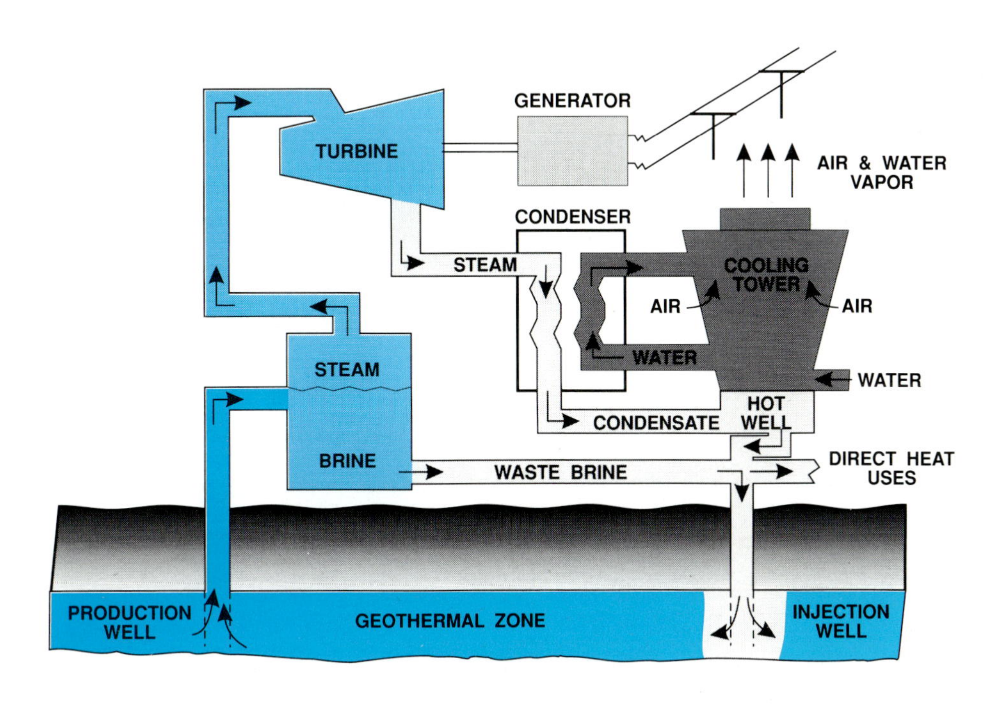

Source: Petroleum Information Corporation, *The Geothermal Resource* (A.C. Nielsen Co., 1979), p. 40.

Figure 4. Schematic of a Double-Flash Geothermal Electricity Generating System for Liquid-Dominated Hydrothermal Resources

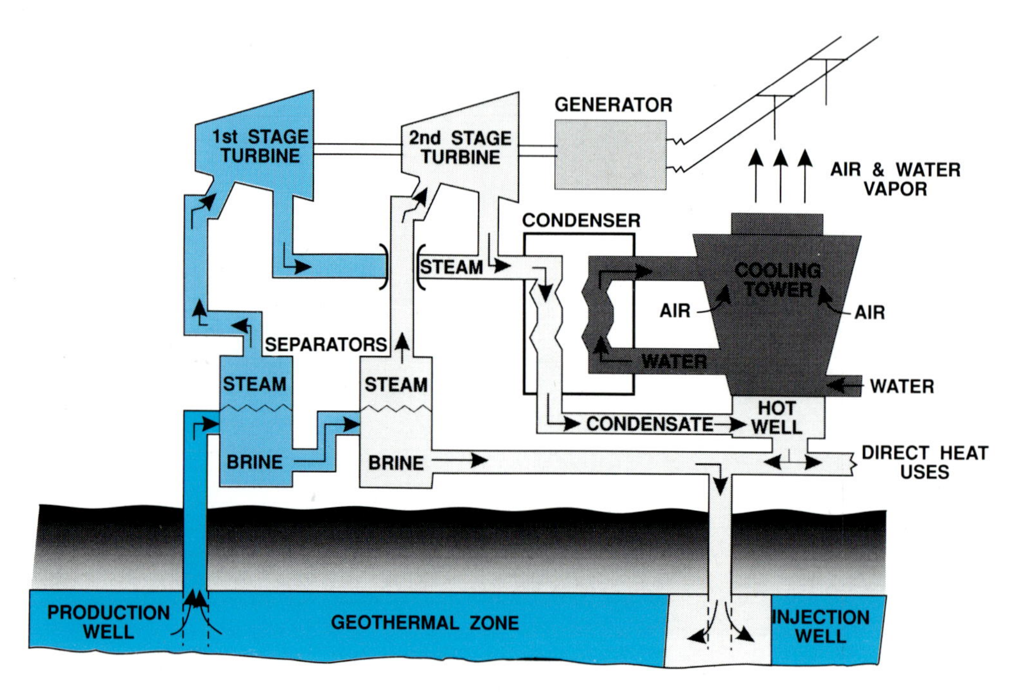

Source: Petroleum Information Corporation, *The Geothermal Resource* (A.C. Nielsen Co., 1979), p. 41.

Figure 5. Schematic of a Binary Geothermal Electricity Generating System for Liquid-Dominated Hydrothermal Resources

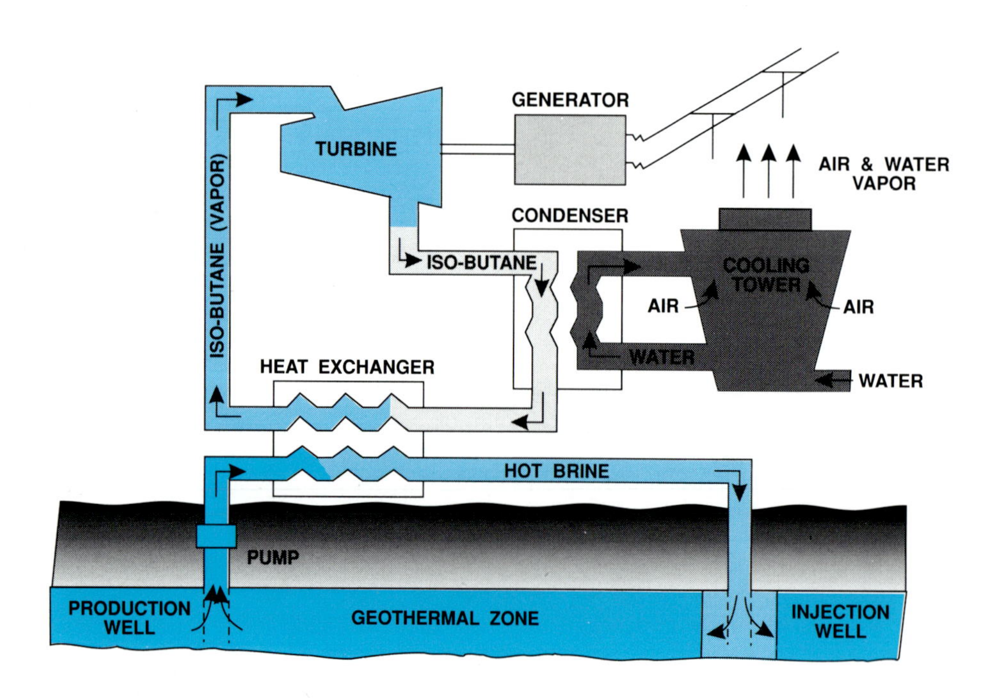

Source: Petroleum Information Corporation, *The Geothermal Resource* (A.C. Nielsen Co., 1979), p. 41.

fluid then rotates a turbine. It is condensed using either cool surface or ground water, or air. After condensing, the working fluid is returned to the heat exchanger. The geothermal fluid is kept in a closed piping system under sufficient pressure to prevent boiling. Binary cycle systems usually are designed to exploit resources with fluid temperatures below 193°C (380°F). Modular plant components are available and can be standardized and prefabricated. Such small-scale, low-cost facilities can be built and installed quickly.

Binary cycle technology has received much attention in recent years. Advantages of binary technology are its capacity to exploit lower temperature fluids that are not economic for flash technologies, minimal atmospheric emissions, and reduced scale and corrosion problems attributable to the closed fluid production system. Disadvantages include the inefficiency and cost of heat exchangers and hazards associated with some working fluids. The development of efficient air-cooled condensers has allowed the use of binary technology in areas with scarce or expensive water supplies at economic energy conversion efficiencies.

Electric Power Supplies

Total Installed Capacity

Following its inception in 1960 and a 20-year period of relative inactivity, the U.S. geothermal industry grew rapidly during the 1980's (Figure 6), with 15.5 billion kilowatthours of electricity being produced in 1989. By the end of 1990, there were 2,719 megawatts of installed capacity at 70 hydrothermal sites (see Appendices C and D and Table 2). About 94 percent of this total was operating in California, and 73 percent of California's total was located at The Geysers in northern California. (A description of The Geysers is presented in the next section of this chapter.) Nevada and Utah had 134 and 32 megawatts of capacity, respectively.

Geothermal plants are generally used to provide baseload generating capacity. Many plants operated

Figure 6. Annual U.S. Hydrothermal Electric Capacity

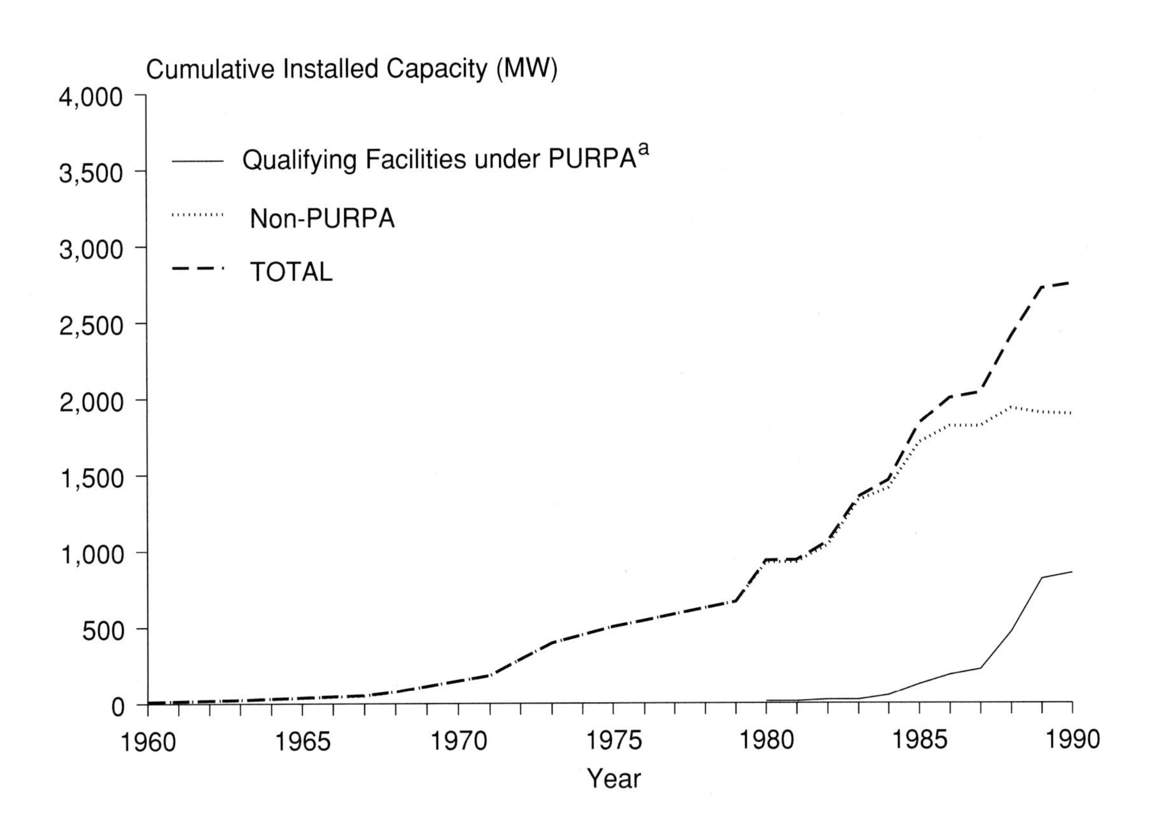

[a]The Public Utility Regulatory Policies Act of 1978 (PURPA) required utilities to offer to buy electrical power from qualifying facilities.
Sources: PURPA—*Geothermal Generating Plants in the United States* (March 1991). Non-PURPA—B. Swezy, Solar Energy Research Institute, personal correspondence to D. DeJarnette, 1991. Total—D. Entingh, Meridian Corporation, personal correspondence, April 9, 1991.

80 to 90 percent of the time, but some were not operational throughout the year.

There has been an increase of more than 80 percent in the installed capacity of hydrothermal plants, from almost 1,500 megawatts in 1984 to 2,719 megawatts in 1990. Approximately one-third of the total installed geothermal capacity at the end of 1990 tapped liquid-dominated resources. This trend toward developing liquid-dominated resources will continue because the unused sites are liquid-dominated.

During the second half of the 1980's, most of the exploratory drilling for geothermal sites was conducted in Nevada, Utah, California, and Oregon. A major find in northern California by UNOCAL in the Glass Mountain-Medicine Lake area in the Cascades is a previously unidentified "hidden" resource.

Although to date more development has occurred in vapor-dominated resources, over 90 percent of known hydrothermal resources are liquid-dominated. In recent years the major additions to geothermal installed capacity have come from liquid-dominated resources. A brief description of each liquid-dominated power plant is provided in Appendix C. A list of geothermal project developers and owners is provided in Appendix D.

The Geysers

The first domestic geothermal power plant was constructed by Pacific Gas and Electric Company at The Geysers (Sonoma County, California) in 1960. When the early geothermal power plants went on line at The Geysers, geothermal steam was marketed directly to Pacific Gas & Electric Company, (PG&E), the local utility company, as a fuel. PG&E built and operated the power plants and purchased steam by contract from the resource developer much as they purchased coal or oil, with some differences. There was a risk that the reservoir would not supply the steam for the contract period at the agreed pressure. Also, the contract price was based on the number of kilowatt-hours of electric power generated by PG&E. The steam supplier contracted to sell his steam at agreed-upon rates at specified pressures. This meant that there was marginal incentive for PG&E to design and build efficient power plants. Even so, geothermal power supplied electricity at competitive prices, with benefit to the utility and the resource developer. By the end of 1990, 26 dry steam power plants were installed with a total capacity of approximately 1,866 megawatts from 439 producing wells.[16]

The Geysers is also the least expensive, most utilized commercially available geothermal resource in the United States. In 1990, The Geysers accounted for more than 11.4 (or 74 percent) of the 15.5 billion kilowatthours (0.16 quadrillion Btu, fossil fuel equivalent Btu[17]) of all domestic U.S. geothermal electrical production.

In 1960, the pressure of the steam resource was nearly 500 pounds per square inch (psi) across the entire reservoir. By 1987, developers announced a reduction in field pressure and the shutting in of some of the wells. At a September 21, 1989, hearing before the California Energy Commission, Pacific Gas and Electric Company, one of the ten users of the field, testified to "a steam shortfall of more than 300 megawatts-electric, or about 22 percent of our installed capacity."[18] By the end of 1990, actual electricity production had fallen as the Pacific Gas and Electric's portion of the production capacity dropped to 1,291 megawatts. Over the years the cumulative effect of extensive steam withdrawal has taken its toll. Today many wells have steam pressures of 200 psi or less. By 2009, electrical generation capacity is projected by PG&E to be 1,025 megawatts, only one-half of current capacity. This situation might be reversed if and when sufficient water is found to recharge the reservoir by injection. Less than 5 percent of the reservoir heat has been extracted to date.

Operational changes have become necessary to efficiently utilize the remaining steam reserves at The Geysers. Two operational strategies have been implemented to conserve these reserves: cycling and load following. Cycling involves planned lowering of the rate of steam extraction during predetermined parts of the day. Cycling reduces the total amount of steam withdrawn while delivering power at peak demand levels. Load following is an efficiency enhancement designed to maximize power plant operation economics by closely matching output with electricity demand. Under a load-following program, plants are throttled back during off-peak hours and returned to high output during peak periods. Both of these strategies allow steam pressure to rebuild and result in an estimated 10 percent increase in available capacity, compared to ordinary baseload operation. Thus, these strategies act

[16]California Department of Conservation, Division of Oil and Gas, *75th Annual Report of the State Oil and Gas Supervisor, 1989* (1990), p. 154.

[17]To allow standardized comparisons between the various fuels, including all renewables and, in this case, geothermal energy, the convention of expressing electricity generation in terms of fossil fuel displacement is used.

[18]U.S. Department of Energy, *Geothermal Progress Monitor*, No. 11, DOE/CE-0283 (Washington, DC, December 1989), p. 23.

to extend the life of the reserves and maximize income by selling at the most advantageous prices.

In an attempt to understand how to respond to the reduction in field pressure at The Geysers, the California Energy Commission appointed a blue-ribbon technical advisory committee to study the problem and make recommendations to correct the situation. In 1990, the committee reported that the reservoir was not being naturally recharged with sufficient groundwater necessary to maintain current steam withdrawals (production levels). Two recommendations were made: (1) artificially recharge the reservoir by injecting surface water, or (2) permanently curtail steam production.[19] California at present is in the midst of a serious drought, and regional water supplies may be too low to sustain substantial injection. For now, the second recommendation is being followed. PG&E predicts a continual decline in productivity until offset by natural recharge or sources of water for injection can be found. Some possibilities for sources of injection fluid are being investigated, including treated sewage effluent.

Costs of Electric Power Generation from Hydrothermal Plants

The future power generation costs for both flash and binary systems are in the competitive range with coal, depending on the location and source of coal. Factors that increase geothermal plant costs include plant siting at moderately problematic reservoirs, and siting at more remote locations.

Estimates of the costs of producing electricity from typical hydrothermal flash and binary plants were developed as part of the background work for the National Energy Strategy. The cost of electricity from a particular geothermal reservoir varies with reservoir temperature, depth, volume, permeability, and a host of other factors. The system depicted here defines an actual plant-size combination in southern California. This combination does not fully represent the full range of variability inherent in hydrothermal systems, and some characteristics, such as the total dissolved solids within the fluid, may not be representative of all hydrothermal plants. Also, cost reductions have occurred since 1986, when the plant characteristics were assessed.

Hydrothermal Flash Plant

Flash system costs estimated in Table 3 assume a utility-owned, 50-megawatt plant in California with the following characteristics:

- Average well depth - 6,000 feet
- Reservoir temperature - 266°C
- Total dissolved solids - 28 percent
- Net brine effectiveness - 7.8 watthour per pound
- Producer average flow - 400,000 pounds per hour
- Capacity factor - 81 percent
- Construction time - 2.5 years.

The resulting capital cost of the system in constant 1990 dollars is about $120 million, or $2,400 per kilowatt, in California, with higher capital costs in other areas of the western United States. The variations are due to differences in the quality of the reservoirs in the different regions. After accounting for operating and maintenance costs, the resulting levelized generation cost is 6.4 cents per kilowatthour.

Table 3. Generation Costs of a 50-Megawatt Hydrothermal Plant in California

	Plant Type	
	Flash	Binary
Capital (million 1990 dollars)		
Discovery	12	14
Well Field	26	28
Power Plant	71	99
Contingency	13	14
Total	**122**	**155**
Fixed Operations and Maintenance (O&M) (dollars per kilowatt per year)		
Well Field	82	61
Power Plant	119	65
Total	**201**	**126**
Variable O&M	0	0
Levelized Generating Costs (cents per kilowatthour)	6.4	6.4

Note: Totals may not equal sum of components due to independent rounding.

Source: Science Applications International Corporation, Inc., *Renewable Energy Technology Characterizations* (Alexandria, VA, September 1990), p. 3.

Hydrothermal Binary Plant

Binary system costs estimated in Table 3 assume a utility-owned, 50-megawatt plant in California with the following characteristics:

- Average well depth - 8,000 feet
- Reservoir saturated temperature - 179°C

[19]"Geysers Watch: Part 1—Plunging Toward 'Abandonment Pressure'," *Geothermal Report*, vol. 19, February 1, 1990, p. 4.

DOE Supporting Research on Problems at The Geysers

With nearly 2,000 megawatts of power generation capacity constructed, the geothermal complex at The Geysers dry steam field in California is the largest in the world. The field has been producing continuously since 1960 and has served an ever-increasing demand since that time. Recently, however, serious problems have developed for which investigations will be required to identify the causes and develop remedial technologies.

Briefly, the problems include a decline in productivity, appearance of corrosive chlorides, increases in noncondensible gases, and the adverse effects of pressure decline on turbine efficiency. More details on the problems reported and DOE's preliminary steps to organize a research strategy may be found in *Geothermal Progress Monitor*, No. 11 (December 1989).

While operations at The Geysers are, and always have been, an industry pursuit, industry has requested government assistance in research aimed at restoring the productivity of the field. The Department of Energy is assisting because The Geysers complex offers an opportunity to devise new technological approaches to managing mature geothermal fields, the first such opportunity in the United States. In addition, the original success of The Geysers, and now its decline, have attracted worldwide attention. Restoration of productive operations at The Geysers will raise industry confidence in the longevity and productivity at all proven and yet-to-be-proven geothermal fields.

An effective program of remedial research at The Geysers is critically dependent on two factors. One is industry's willingness to share existing data about the characteristics of the field. The second is industry cost-sharing. Funding available from DOE (over $1 million in FY 1990) is not, by itself, sufficient to solve the problems.

Despite the fact that The Geysers may have been the subject of more study than any other geothermal reservoir, several parameters remain poorly understood. These include the initial distribution and amount of liquid water, reservoir thickness, matrix permeability, and characteristics of the fracture network.

In order to fill these information gaps, to address the other problems plaguing operations at The Geysers, and to determine whether water injection is the optimum "cure" for The Geysers, the DOE funded 11 research projects for FY 1990. The geochemical research projects include:

- A thermodynamic investigation of hydrogen chloride in steam by the Oak Ridge National Laboratory
- Development of new vapor phase tracers by the University of Utah Research Institute (UURI) that can be used to quantify the mass recovery of injected fluids
- A study of steam chemistry by the U.S. Geological Survey (USGS) with the cooperation of operating companies and the International Institute for Geothermal Research
- Fabrication of a six-liter downhole fluid and gas sampler by Lawrence Berkeley Laboratory (LBL) based on a smaller version used successfully in the Imperial Valley and in a Continental Scientific Drilling Project well in the Valles Caldera, New Mexico.

(continued on next page)

The geophysical research projects include:

- Microearthquake studies at The Geysers by LBL in conjunction with the Coldwater Creek Operator Corporation using a 16-station array presently in place in the northwest portion of the field

- Continuation of the ongoing Lawrence Livermore National Laboratory seismic attenuation study to locate steam.

The reservoir engineering projects include:

- Investigation of the phenomenon of water adsorption in porous rocks at Stanford University through lab experiments with Geysers core material (in parallel, engineering methods for using adsorption to plan development and forecast results will be explored)

- Examination by Stanford of all the tritium survey data collected by several operating companies in light of physical information on the wells (such as feed point depth) as well as subsequent performance (e.g., temperature and pressure).

The reservoir modeling projects include:

- Development by LBL of a data base for The Geysers, incorporating all available geological, geochemical, and reservoir engineering data, to be subjected to theoretical and applied studies to quantify the impact of increased injection

- Documentation by LBL of several of the MULKOM model's fluid property modules so that the simulation capabilities of the code can be made available to the public.

The only geological research project funded so far involves fluid inclusion studies by UURI on Geysers core samples where the age relationships among the secondary minerals can be defined. Results of this initial work will provide the necessary background for interpreting similar data to be obtained from cuttings.

Source: Modified from U.S. Department of Energy, *Geothermal Progress Monitor*, No. 12 (December 1990), p. 5.

- Total dissolved solids - 6 percent
- Net brine effectiveness - 6.2 watthours per pound
- Producer average flow - 510,000 pounds per hour
- Capacity factor - 81 percent
- Construction time - 2.5 years.

The resulting capital cost of the system in constant 1990 dollars is about $155 million, or $3,100 per kilowatt, in California, with higher costs in remote regions. After accounting for operating and maintenance costs, the levelized generation costs are estimated at 6.4 cents per kilowatthour.

Market Development Trends

A major change in marketing strategy now faces the developers of geothermal power. The reinterpretation of PURPA by a number of State regulatory commissions requires independent power producers to bid competitively for providing new electric generating capacity, rather than utilities having to purchase power at full avoided costs. This provides market access for geothermal developers who can produce power at prices competitive with other resources, taking into account the full life cycle costs for new power. In two competitive bid cycles, Sierra Pacific (Nevada) awarded 8 out of 11 contracts to geothermal projects. Los Angeles Department of Water and Power (California) recently asked for competitive bids for power and received several geothermal power project offers at competitive prices. Puget Power (Washington) awarded one of three competitively bid power sales contracts to a geothermal project.

A problem facing geothermal developers has been the sluggish growth in power usage experienced in the West. Most utilities in that region expect to have capacity surpluses for the next 3-5 years. Some western utilities currently have excess power available for sale.[20]

Lack of power markets has resulted in a slowdown of exploration activity (Figure 7). Over 95 percent of current geothermal drilling activity is developmental rather than exploratory.[21] Exploration does continue along New Mexico's Rio Grande rift and in Oregon's and California's Cascade mountain range.

Public Utility Regulatory Policies Act

A milestone for the geothermal industry occurred in 1978, when Congress enacted the Public Utility Regulatory Policies Act (PURPA) to encourage the development of small-scale independent electricity production, cogeneration, and energy conservation. In 1980, the Federal Energy Regulatory Commission (FERC) produced implementation guidelines which required utilities to offer to buy power from "qualifying facilities." The mandated "offer to buy" effectively eliminated price competition among producers by establishing price supports for the qualifying facilities (QF). QFs are those which derive at least 75 percent of their fuel from renewable resources and are not more than 50 percent owned by a utility or utility subsidiary.

Figure 7. Exploratory Geothermal Energy Wells in California, 1976-1989

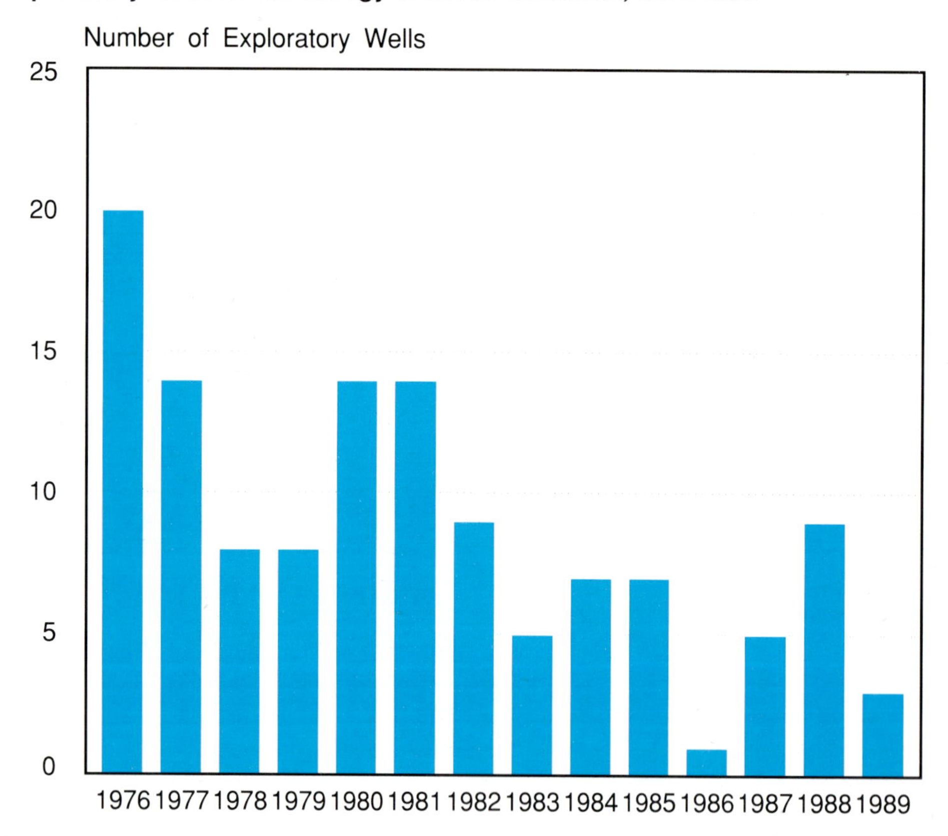

Source: California Department of Conservation, Division of Oil and Gas, *75th Annual Report of the State Oil and Gas Supervisor, 1989* (1990), pp. 154-156.

[20]Williams, S., and Porter, K., *Power Plays: Geothermal* (Investor Responsibility Research Center, 1989), pp. 168-169.

[21]Williams, S., and Porter, K., *Power Plays: Geothermal* (Investor Responsibility Research Center, 1989), pp. 165-181.

The purchase price of the power is stipulated by FERC regulations as each utility company's full avoided cost.[22] Full avoided costs are those costs which a utility would have incurred had it generated the power at its own facilities. The full life cycle cost of generating new power was not considered in the avoided cost formulas.

FERC regulations also tied the price paid for power from QFs to the price of existing fuels. Although the cost of old, large-scale hydropower, coal and nuclear fuel was low, the cost of oil and natural gas was high, elevated by Federal regulation of natural gas prices and OPEC manipulation of the oil market. This made geothermal power competitive even though hydrothermal technology was still evolving.

Standard Offer Number Four Contracts

Even though PURPA is a Federal statute, it is implemented by State utility commissions.[23] In 1980, the California Public Utilities Commission developed a series of standard contracts designed to govern the terms and conditions of power sales by QFs.[24] One of that series, known as Standard Offer Number Four (SONF), materially enhanced the development of geothermal resources in California and Nevada.[25] These 30-year contracts provided a developer with a price set at the utility's avoided cost. Time limits for the completion of projects and the timely delivery of prescribed amounts of power were specified in the contract. These lucrative contracts were available until the anticipated shortfall of power was met by sufficient signed contracts in 1986. At that point, industry efforts shifted from entering into new contracts to meeting existing contract delivery dates.

The SONF contracts gave geothermal resource developers an opportunity for marketing electric power which allowed them to both develop the resource and produce the power. These contracts guaranteed independent power producers a ready market for predetermined quantities of power. However, with the drop in oil prices and the reassessed projections of power supply after 1985, utilities ceased entering into SONF contracts. While most geothermal SONFs were delivered on time, a few SONF contracts were terminated due to non-delivery. This occurred for several reasons: (1) Project economics for some marginally economic resources, particularly those tapping high salinity brines in the Imperial Valley area, were determined to be uneconomic, even at SONF prices; (2) complex financing arrangements fell through due to investor uncertainty about long-term resource economics; and (3) oil companies supporting geothermal development on oil revenues had less capital available for geothermal development projects.

Although in 1989 the tax credits for geothermal development expired, these credits have now been extended, which may reassure some investors and encourage geothermal development.[26]

Planned Additions

The planned additions to geothermal power generation (Table 4) reflect the short planning horizons of the utilities and States. Since hydrothermal power plants can be constructed and brought on-line a year or two after the award of a power contract, information on additions beyond 1995 is not available, except in a general sense.

Currently, power contracts are in place to bring another 278 megawatts of geothermal capacity into operation by the end of 1995, principally in California and Nevada. An additional 386 megawatts of geothermal capacity are in the planning stages. If water is available for injection at The Geysers, 346 megawatts of additional power is planned.

In the State of Hawaii, the growth follows a decade of research and development, funded by industry and government. The Puna Geothermal Venture (Ormat Energy Systems) has a contract with the Hawaii Electric Light Company to provide 27 megawatts of power in 1991.[27]

In Utah, one plant with 7 megawatts capacity is planned to begin generating power in 1991; two more

[22]"Overview Report—New Policies Needed Now for the 1990's—Part 1: The California Perspective," *Geothermal Report*, vol. 18, June 15, 1989, p. 4.

[23]Williams, S., and Porter, K., *Power Plays: Geothermal* (Investor Responsibility Research Center, 1989), p. 10.

[24]"Overview Report—New Policies Needed Now for the 1990's—Part 1: The California Perspective," *Geothermal Report*, vol. 18, June 15, 1989, pp. 4-5.

[25]Northwest Power Planning Council, *Assessment of Geothermal Resources for Electric Generation in the Pacific Northwest* (Portland, OR, 1989).

[26]Williams, S., and Porter, K., *Power Plays: Geothermal* (Investor Responsibility Research Center, 1989), pp. 165-181.

[27]G.O. Lesperance, State of Hawaii Department of Business, Economic Development and Tourism, Energy Division, personal correspondence to H. Walton, February 4, 1991.

plants might possibly come on-line by 1995, with a total of 45 megawatts generating capacity. In Nevada, three plants with a total of 101 megawatts capacity are planned by 1995, with one more plant with 34 megawatts capacity yet to have a power contact. Oregon has one planned addition, a 10-megawatt plant, which might have a power contract prior to 1995.

Table 4. Hydrothermal Power Plants: Utility Generating Capacity Planned Additions and Announced Projects, 1991-1995

Year of Planned Service	State	Name of Plant	Capacity (Megawatts)	Power Contract Awarded: Yes	Power Contract Awarded: No
1991	CA	QF Geothermal 91	[a]3	not available	
1991	CA	Non-Utility Geothermal	[a]21	not available	
1991	HI	Puna	[b]27	X	
1991	UT	Cove Fort 1	[a]7	not available	
1993	HI	Island of Hawaii	[b]25		X
1995	CA	Surprise Valley	[c]10	X	
1995	CA	Clear Lake	[c]50		X
1995	CA	Coso	[c]100	X	
1995	CA	Long Valley-Low	[c]20		X
1995	CA	Randsburg	[c]10		X
1995	CA	Salton Sea	[c]61		X
1995	CA	Geysers	[c,d]346	[d]X	
1995	CA	Brawley	[c]50		X
1995	CA	East Mesa	[c]40	X	
1995	CA	Medicine Lake	[c]50		X
1995	NV	Beowawe	[c]34		X
1995	NV	Steamboat	[c]34	X	
1995	NV	Dixie Valley	[c]38	X	
1995	NV	Desert Peak	[c]29	X	
1995	OR	Alvord Desert	[c]10		X
1995	UT	Roosevelt	[c]16		X
1995	UT	Cove Fort	[c]29		X

Note: Potential projects with no power contracts are highly uncertain.

Sources: Compiled by the Office of Coal, Nuclear, Electric and Alternate Fuels:

[a]NERC 1989 Electricity Supply and Demand for 1990-2000.

[b]G.O. Lesperance, State of Hawaii Department of Business, Economic Development and Tourism, Energy Division, personal correspondence to H. Walton, February 4, 1991.

[c]Susan Petty Consulting, Inc., oral communication, to T. Burski, April 15, 1990.

[d]The capacity already exists. The additional power is based on an assumed increase in availability of water for the ongoing injection process, and is highly uncertain.

3. Electric Power Generation Potential

Introduction

In this chapter, recent assessments[28] of the potential for generating electricity with geothermal energy are reviewed. Projections of geothermal development potential were developed in 1990 by the Interlaboratory Task Force on Renewable Energy,[29] under the direction of the Solar Energy Research Institute (SERI), as background information for use in developing DOE's National Energy Strategy (NES). These projections were reviewed by the Renewable Energy Technologies Subgroup of the National Energy Modeling System, composed of the following organizations: the Energy Information Administration, the Office of Conservation and Renewable Energy, Oak Ridge National Laboratory, Sandia National Laboratory, the Solar Energy Research Institute, Meridian Corporation, and Science Applications International Corporation. This review resulted in a revised set of projections used by EIA for the *Annual Energy Outlook 1991 (AEO)*.[30]

Finally, in a follow-on effort to the NES, the EIA and DOE's Office of Conservation and Renewable Energy jointly commissioned a study of potential hydrothermal electric energy resources. The study on the supply of geothermal power from hydrothermal resources, conducted through the Sandia National Laboratory, attempted to expand upon the resource assessments previously conducted by groups such as USGS, and it represents the most recent quantitative effort to evaluate the magnitude of U.S. geothermal resources and the associated costs of electric generation.

Interlaboratory Task Force on Renewable Energy Assessment

The Interlaboratory Task Force[31] developed projections of electricity power supply from geothermal resources to 2030, based on consensus judgments on expected geothermal technology costs and penetration rates. Mid-range projections were constructed for each of three technology deployment scenarios: (1) a Business-As-Usual Scenario, reflecting current levels of research activity; (2) an enhanced level of Government support (Research, Development and Demonstration Intensification Scenario); and (3) a renewable subsidy case, in which geothermal electricity was awarded a premium of 2 cents per kilowatthour (National Premiums Scenario) to reflect its benefit as a clean energy source in competition with fossil-fired and nuclear-powered electricity generation.

To derive its projections, the Interlaboratory Task Force employed a "competitive market construct" of forecasted energy consumption levels and prices. Projections from the EIA's *Annual Energy Outlook 1989* and *Annual Outlook for U.S. Electric Power 1989* were used to define the market construct to the year 2000, while a preliminary run of the FOSSIL2 model of DOE's Office of Policy, Planning and Analysis extended the construct out to the year 2030. Cost curves for each geothermal technology were then developed based on preliminary assessments of technology cost and performance improvements expected over the forty-year study period. The cost curves were applied against a set of regionalized market demands to derive the forecasted time-line for technology market penetration and the associated energy supply levels.

[28]Earlier Federal assessments not covered herein include a 1974 report by the Project Independence Task Force, and projections of generation potential made in 1979 by both the USGS and National Research Council.

[29]U.S. Department of Energy, *The Potential of Renewable Energy*, An Interlaboratory White Paper, SERI/TP-260-3674, (Solar Energy Research Institute, Golden, CO, March 1990), p. 1.

[30]Energy Information Administration, *Annual Energy Outlook 1991*, DOE/EIA-0383(91) (Washington, DC, 1991), Table A6, p. 50.

[31]U.S. Department of Energy, *The Potential of Renewable Energy*, An Interlaboratory White Paper, SERI/TP-260-3674 (Solar Energy Research Institute, Golden, CO, March 1990), p. 2.

Projections were made for each of the four candidate geothermal technologies (Table 5). In the Business-As-Usual Scenario, the Interlaboratory Task Force projected a slow growth of geothermal electricity generation to 0.5 quadrillion Btu by 2010. Under the assumptions of an 80-percent capacity factor and a fossil fuel-equivalent heat rate of 10,235 Btu per kilowatthour,[32] this converts to about 6,000 megawatts of capacity. Most of the expansion is in the use of hydrothermal resources, although hot dry rock contributes a substantial share. By 2030, generation is projected to grow to almost 1.0 quadrillion Btu, or an annual capacity of 10,600 megawatts.

Renewable Energy Technologies Subgroup Assessment

In 1990 and 1991, the Department of Energy published two documents forecasting the penetration of hydrothermal technology into the electric utility sector: the *Renewable Energy Excursion*[33] and the *Annual Energy Outlook 1991 (AEO)*.[34] The analysis undertaken by the Renewable Energy Technologies Subgroup for the development of the National Energy Strategy (NES) examined several potential energy supply and demand scenarios over the period 1990 through 2030.

The two fundamental questions or cases addressed by the studies for the NES analysis were:

- **Baseline Case:** How much energy might be derived from hydrothermal sources by 2030, under existing laws and regulations?
- **Excursion:** By how much might that contribution be reasonably increased through accelerated improvements in cost and performance of hydrothermal technologies?

The Excursion did not explicitly consider tax credits or other financial incentives for hydrothermal or factors that would change the costs of conventional fuels. Improvements in hydrothermal technology costs and performance were assumed to result from research and development efforts, industry investment, or technological breakthroughs.

The fundamental modeling tool for integrated analysis in the NES development effort was the FOSSIL2 Model. FOSSIL2 provided a large-scale dynamic simulation of U.S. energy supply and demand over a 40-year time period. The model integrated new technologies into the model. Technologies that are more economical than others were assigned correspondingly larger market shares. Technologies slightly more economical than others received only slightly larger market shares, with

Table 5. Interlaboratory Projections of Geothermal Electric Capacity and Generation, 1988-2030 [a]
(Quadrillion Btu)

	1988		2000		2010		2020		2030	
	Capacity[b] (MW)	Generation (Quads)[c]	Capacity[b] (MW)	Generation (Quads)[c]	Capacity[b] (MW)	Generation (Quads)[c]	Capacity[b] (MW)	Generation (Quads)[c]	Capacity[b] (MW)	Generation (Quads)[c]
Hydrothermal	2,565	0.23	3,234	0.29	5,242	0.47	6,581	0.59	6,246	0.56
Geopressured	--	--	--	--	335	0.03	558	0.05	781	0.07
Hot Dry Rock	--	--	112	0.01	223	0.02	1,227	0.11	3,011	0.27
Magma	--	--	112	0.01	112	0.01	223	0.02	558	0.05
Total	**2,565**	**0.23**	**3,458**	**0.31**	**5,912**	**0.53**	**8,589**	**0.76**	**10,596**	**0.95**

[a]U.S. Department of Energy, *The Potential of Renewable Energy*, An Interlaboratory White Paper, SERI/TP-260-3674 (Solar Energy Research Institute, Golden, CO, March 1990), Business-As-Usual (BAU) case, Table C-1, p. C-7.
[b]Electric generating capacity is based on fossil fuel equivalent heat rate of 10,235 Btu per kilowatthour. Source: Energy Information Administration, Office of Coal, Nuclear, Electric and Alternate Fuels.
[c]Quadrillion Btu.

[32]To allow standardized comparisons between the various fuels, including all renewables and, in this case, geothermal energy, the convention of expressing electricity generation in terms of fossil fuel displacement is used. That is, geothermal energy is credited at the fossil fuel rate of about 10, 235 Btu per kilowatthour; the heat equivalent of electricity is 3,412 Btu per kilowatthour; The Geysers heat rate is estimated at approximately 21,096 Btu per kilowatthour.

[33]Energy Information Administration, *Renewable Energy Excursion: Supporting Analysis for the National Energy Strategy*, SR/NES/09-04 (Washington, DC, December 1990), Table 2, p. 22.

[34]Energy Information Administration, *Annual Energy Outlook 1991*, DOE/EIA-0383(91) (Washington, DC, 1991), Table A6, p. 50.

the market share increasing exponentially for the preferred technologies.

For the electric utility sector, FOSSIL2 projected: (1) the new capacity that might be built in response to future demand, (2) how existing and new capacity might be dispatched to satisfy anticipated demand, and (3) electricity rates in accordance with utility ratemaking procedures. The electricity prices were fed back to the demand sectors to determine current and future load growth. Utility and nonutility grid-connected generating units competed for a share of the new capacity on a least life-cycle cost basis, using levelized costs per kilowatthour for comparing alternative technologies used by utility companies. Utility avoided costs was the measure used for alternative technologies used by nonutilities. The levelized costs included capital costs, operating and maintenance costs, and fuel costs converted to a cost-per-kilowatthour based on design capacity factors.

The impacts of geothermal resource development presented in the *AEO*[35] are based on the analysis undertaken for the NES, and are consistent with the NES projections through the year 2010 (Table 6). The projections for the *AEO* indicate an expansion of hydrothermal facilities to 10,650 megawatts by 2010, resulting in electric generation of 0.82 quadrillion Btu. A major portion of this expansion could be expected from fields developed in California other than The Geysers. Additionally, there would be substantial expansion in other States, including Oregon, New Mexico, Nevada, and Utah. Assessments for the DOE's Office of Conservation and Renewable Energy have indicated that continued Federal R&D funding will be required to achieve any of the NES estimates of geothermal electricity production.

Sandia National Laboratory Assessment

The primary objective of the Sandia hydrothermal assessment was to expand on existing knowledge concerning the resource base to determine the available supply of electric power from geothermal resources and the cost of producing that power in 20 years (2010) and in 40 years (2030)—so-called "supply curves." The hydrothermal resources of Arizona, California, Colorado, Hawaii, Idaho, New Mexico, Nevada, Utah, and Washington were examined in the study. The resources of Alaska were not considered because of their remote location. This assessment was based on estimates of the quantity of energy that could physically be available for sale over a 30-year reservoir life, not on the quantity that would be sold. Considerations that could constrain development, such as environmental issues, proximity to power transmission lines, the local power market, and social and political considerations were all outside the scope of the study. The estimates are limited by exploration technology, the availability of exploration equipment and infrastructure, and the rate at which this type of exploration could proceed.

Detailed descriptions of the results, the methodology, and the electric power cost assumptions used in the study are provided in Appendix B of this report.

Assuming the use of current technology for currently identified resources that could produce electricity at the busbar for 6 cents per kilowatthour, 6,000 megawatts could become available in 20 years, and 14,000 megawatts could become available in 40 years. Under the Base Case assumptions, at 12.5 cents per kilowatthour of delivered electricity at the busbar,

Table 6. Projections for Geothermal Capacity and Generation, 1990-2030

	1990	1995	2000	2005	2010	2020	2030
Capacity (Megawatts)	2,590	3,250	6,250	9,650	10,650	19,500	23,400
Generation							
(Billion Kilowatthours)	16	21	43	71	79	143	172
(Quadrillion Btu)	0.17	0.21	0.45	0.73	0.82	1.50	1.80

Sources: Energy Information Administration, *Annual Energy Outlook 1991*, DOE/EIA-0383(91) (Washington, DC, 1991), Reference Case, Table A6, p. 50.; *Renewable Energy Excursion: Supporting Analysis for the National Energy Strategy*, SR/NES/90-04 (Washington, DC, December 1990), Baseline Case (2020, 2030), Table 2, p. 22.

[35]Energy Information Administration, *Annual Energy Outlook 1991*, DOE/EIA-0383(91) (Washington, DC, 1991), Table A6, p. 50.

Table 7. Comparison of Estimates of Potentially Installable Capacity from U.S. Hydrothermal Electricity Resources
(Megawatts)

	Installed and Potentially Installable Capacity			Resource Base
	1990	2010	2030	
Installed	2,719			
Interlaboratory Task Force Projections[a]		5,900	10,600	
Renewable Energy Subgroup Assessment[b]		10,650	23,400	
Sandia Resource Assessment (6 Cents per Kilowatthour)[c,d]				
Current Technology (Base Case), Identified Resources		6,000	14,000	
Improved Technology, Identified Resources		8,000	18,000	
Current Technology, Identified and Estimated Unidentified Resources		8,500	19,000	
Improved Technology, Identified and Estimated Unidentified Resources		11,000	31,000	
Sandia Resource Assessment (12.5 Cents per Kilowatthour)[d,e]				
Current Technology (Base Case), Identified Resources		10,000	22,000	
Improved Technology, Identified Resources		10,500	24,500	
Current Technology, Identified and Estimated Unidentified Resources		14,000	36,000	
Improved Technology, Identified and Estimated Unidentified Resources		15,500	44,000	
USGS Resource Assessment[d,f]				
Identified Resources	NA[g]	NA[g]	NA[g]	23,000[h]
Identified Plus Undiscovered Resources	NA[g]	NA[g]	NA[g]	95,000-150,000[h]
BPA Resource Assessment[d,i]				
Identified Plus Undiscovered Resources	NA[g]	NA[g]	NA[g]	185,000-280,000[h]

Sources: Compiled by the Office of Coal, Nuclear, Electric and Alternate Fuels:
[a]U.S. Department of Energy, *The Potential of Renewable Energy*, An Interlaboratory White Paper, SERI/TP-260-3674 (Solar Energy Research Institute, Golden, CO, March 1990), Business As Usual Case, Table C-1, p. C-7. The estimates for 2010 and 2030 include 670 megawatts and 4,710 megawatts, respectively, of non-hydrothermal resources (i.e., hot dry rock, geopressurized, and magma).
[b]Energy Information Administration, *Annual Energy Outlook 1991*, DOE/EIA-0383(91) (Washington, DC, 1991), Baseline Case, Table A6, p. 50. In 2010, the 10,650 megawatts were estimated to be competitive with other energy sources at less than 6 cents per kilowatthour for delivered electricity.
[c]Petty, S., Livesay, B.J., and Geyer, J., *Supply of Geothermal Power from Hydrothermal Sources: A Study of the Cost of Power Over Time*, prepared for the U.S. Department of Energy (Sandia National Laboratory, 1991) (Draft), Tables 1 and 2. Capacity was estimated to be supplied at a busbar cost of 6 cents per kilowatthour. See Appendix B.
[d]Market factors are not considered.
[e]Petty, S., Livesay, B.J., and Geyer, J., *Supply of Geothermal Power from Hydrothermal Sources: A Study of the Cost of Power Over Time*, prepared for the U.S. Department of Energy (Sandia National Laboratory, 1991) (Draft), Tables 1 and 2. Capacity was estimated to be supplied at a busbar cost of 12.5 cents per kilowatthour. See Appendix B.
[f]Muffler, L.J.P., editor, *Assessment of Geothermal Resources of the United States—1978*, U.S. Geological Survey Circular 790 (1978), Table 4, p. 41. Excludes reservoirs in Cascades and National Parks.
[g]Not applicable. The estimates of recoverable energy are unbounded by time.
[h]Estimates of recoverable energy unbounded by time. Data represent annual electrical generation capacity potentially obtainable over a 30-year lifespan of the resources.
[i]Bloomquist, R.G., Black, G.L., Parker, D.S., Sifford, A., Simpson, S.J.H., and Street, L.V., *Evaluation and Ranking of Geothermal Resources for Electrical Offset in Idaho, Montana, Oregon and Washington*, Bonneville Power Administration Report, DOE/BP13609, Volume 1, Table 3.2, p. 75. Includes National Parks and wilderness areas.

almost 10,000 megawatts could become available in 20 years and 22,000 megawatts in 40 years (Table 7). In the most optimistic scenario, 15,500 megawatts could be available in 20 years and 44,000 megawatts in 40 years.

A Comparison of Projections

A comparison between the Sandia estimates of the availability of electric power supplies and USGS and BPA resource estimates shows a wide range of potential for utilization (Table 7). These estimates vary depending on assumptions concerning the availability of knowledge concerning the resource base, the portion of the resource base used in the estimates, and subjective assessments of the technology improvements that can be achieved in the future. The USGS and BPA resource assessments are described in Appendix A, and current installed capacities are reported in Chapter 2.

Estimates that are economically constrained by busbar costs should be distinguished from those that were made without economic considerations. The Sandia study estimated the level of resources available for electric power generation at 6 cents per kilowatthour in 20 years to be between 6,000 and 11,000 megawatts, at the busbar. This compares to 2,719 megawatts of currently installed capacity. The total quantity of electric power estimated by Sandia to be available at 12.5 cents per kilowatthour at the busbar in 20 years is roughly 60 percent of the USGS assessment for identified high-temperature hydrothermal resources. The most optimistic projection of electric power available from hydrothermal resources in 40 years estimated at 44,000 megawatts in the Sandia study is approximately one-fourth to one-sixth the amount of total identified and undiscovered hydrothermal resources estimated by the BPA. The BPA estimated that between 185,000 and 280,000 megawatts of potentially installable capacity could be made available when no economic and market factors are considered.

Considerations Affecting Future Utilization of Geothermal Resources

The prospects for increased exploitation of hydrothermal resources are dictated by the conditions applicable to any existing energy technology: access to secure, long-term energy supplies with known environmental impacts and predictable costs; the maintenance of capital and operating costs at a level that produces competitively-priced energy relative to other energy technologies; and overcoming political and institutional barriers that destabilize the business climate. The conditions of utilization are highly dependent on the continued evolution of effective technology, energy resource management, and regulation of utilities. The probability of meeting many of these conditions is uncertain.

Numerous factors, both positive and negative, affect the development of electric power generation from hydrothermal resources (see the following box). This section considers three areas with conditions which will influence the future development of geothermal resources. First, several considerations associated with continued exploitation of hydrothermal resources are discussed, followed by a discussion of operational and market penetration factors applicable to commercially viable geothermal technologies. The chapter concludes with a very brief review of technical uncertainties surrounding the long-term exploitation of magma, geopressured, and hot dry rock resources.

Environmental Considerations

Environmental issues provide a significant impetus to the development of geothermal resources. Hydrothermal geothermal technology is relatively "clean" with minimal adverse impact on the environment.[36] Since geothermal development entails no combustion, its atmospheric emissions are limited to the dissolved gases that are released during depressurization in open cycle systems. Carbon dioxide is released in direct steam and flash systems at a typical rate of 55.5 metric tons per gigawatthour, or at approximately 11 percent of the rate of gas-fired steam electric plants. The amount will vary from site to site. Some recent plants, particularly those at Coso Hot Springs, California, inject the noncondensible gases, limiting emission of greenhouse gases to well testing and unplanned plant outages. Some hydrothermal development employs lower temperature, binary cycle technology. Carbon dioxide emissions from such closed cycle systems is negligible. Similarly, some analysts project that most of the prospective long-term geothermal potential will be derived from exploiting hot dry rock and magma resources, and these technologies will not entail any significant emissions of carbon dioxide.

For electric power generation, geothermal energy could potentially be utilized to displace conventional baseload

[36]U.S. Department of Energy, *Geothermal Progress Monitor*, No. 12 (December 1990), pp. 15-16.

Factors Affecting the Hydrothermal Electricity Generation Industry

ACCESS TO ENERGY SUPPLIES

National Security

- The indigenous nature of the resources make the hydrothermal power supply immune to the effects of foreign governments' oil policies

Resource Availability

- Known reserves are leased or owned by companies poised to develop the resources when profits are assured

- Identified resources provide assurances that hydrothermal energy supplies will be available for the foreseeable future

Environmental Effects

- Hydrothermal energy technologies have a significant advantage over conventional power generating technologies in air pollution impacts, hazardous waste generation, water use, water pollution, and carbon dioxide emissions

- The benign nature of the hydrothermal power plants, relative to fossil fuel plants, has become more important since the recent Clean Air Act legislation was signed into law

- Pollution credits under acid rain legislation may be available

- Site restoration and decommissioning costs have not been incorporated fully with this and all other technologies

- Environmental impact evaluation in the permitting process, when compared to other energy sources, should show hydrothermal to have additional long-range benefits

COMPETITIVE PRICING

Reserve Capacity

- Excess reserve capacity exists in most regions with hydrothermal resources

Competitive Bidding

- Fossil fuel prices affect profitability (high prices mean geothermal is more competitive)

- Competition (natural gas plants, imported hydropower) may have lower busbar and delivered costs. Others (coal, oil) may have equal or higher costs, depending on the extent of the incorporation of environmental costs

- The cost of generating power from geothermal resources can be fairly stable over extended periods of time

(continued on next page)

Financial

- The short lead times for construction reduce the financing costs
- Modular design reduces uncertainties through standardization
- High front-end costs deter investors
  - Exploratory drilling is costly, especially when compared to the risk involved in exploratory drilling for oil
  - Trouble-related drilling costs are uncertain
  - Pilot plants are often required, increasing front end costs and risk
- Expected operational reservoir life has not been reached at current production sites, creating conditioned expectations for others and affecting investor activities
  - Heat production behavior over lifetime of the reservoir affects efficiency projections
  - Flow rates of hot fluid may decrease over time
  - Corrosion by highly saline liquids and hydrogen chloride has occurred at a few sites
- Funding and approval of transmission access lines is conditioned on generally just one utility's cooperation

POLITICAL AND INSTITUTIONAL BARRIERS

Local Populace Reactions

- Conflict with local beliefs or traditions, as in Hawaii, may affect development
- Roads and transmission lines which disturb the environment may affect the permitting process
- Drilling noise may require sound abatement equipment or restrict hours of operation
- Gases being released may require pretreatment

Regulation

- Reform of the Public Utility Holding Company Act (PUHCA) may afford additional opportunities
- National park areas contain unavailable resources, which limits development of geothermal power
- Regulatory movement toward least-cost planning with inclusion of externalities could encourage development

Source: Compiled by the Office of Coal, Nuclear, Electric and Alternate Fuels.

generation. Baseload power requirements in the western United States are currently met with coal, nuclear, hydroelectric, natural gas, and hydrothermal capacity. Coal appears to constitute the dominant fuel of economic choice for new baseload generation capacity in coming decades. The principal atmospheric emission problems caused by current coal-based electricity generation facilities are acidic precipitation and potential climate change from emissions of greenhouse gases. Geothermal resources offer an attractive alternative to ameliorate these atmospheric emissions.

Environmental issues that could adversely affect the future development of geothermal resources include water requirements, air quality issues, waste effluent disposal, subsidence, noise pollution, and location issues.

Water Requirements

Some geothermal power plants use large quantities of cooling water.[37] For example, a 50-megawatt water-cooled binary plant requires more than 5 million gallons of cooling water per day (100,000 gallons per megawatt per day). This is significant since many geothermal resources are located in arid regions where water is a scarce and regulated commodity. Thus, long-term access to sufficient quantities of cooling water could be an important constraint in the planning phase of development at some locations. However, plants can use dry cooling systems at a small increase in capital cost and some net output loss during the summer. If other aspects of project economics are good, these plants can be used even where water is not available. Flash steam plants can also have a substantial portion of their water needs supplied by steam condensate, and if the residual geothermal liquid is fresh enough, it can be used as well.

At The Geysers in California, production declines could be substantially improved by injection of water from external sources. However, there is a shortage of water for recharging the hydrothermal reservoir, due to a multiyear drought. Competition with rural farmers and urban residences for water has led to shut-in capacity rather than a recharging of the aquifer. Use of treated sewage effluent could supply needed recharge if this were economically feasible.

Air Quality

There are no air emissions where closed-loop binary technology is used since the system does not allow exposure of the hydrothermal fluid to the atmosphere. Naturally occurring chemical compounds may be released into the atmosphere as a byproduct of the extraction of geothermal energy at some sites.[38] The emissions can include varying concentrations of hydrogen sulfide, hydrogen chloride, carbon dioxide, methane, ammonia, arsenic, boron, mercury, and radon. The concentrations of emissions vary from site to site depending on resource characteristics and the technology applied.

Emission of hydrogen sulfide is often a concern at steam and flash plants because it exhibits a characteristic "rotten egg" odor at low concentrations. At high concentrations, it is toxic. Air quality standards can be inexpensively achieved by installing hydrogen sulfide abatement systems that range in cost between 0.1 and 0.2 cents per kilowatthour. Noncondensible gas emissions such as carbon dioxide and hydrogen sulfide can be reduced by reinjection into the reservoir. However, the long-term effects of this practice on the geothermal reservoir remain unknown.

Waste Effluent Disposal

To date, all waste streams from geothermal facilities have satisfied California standards through either treatment or emission control. Research efforts designed to alleviate disposal problems continue. However, geothermal fluids can contain large quantities of dissolved solids, such as at the Salton Sea field in California. The energy extraction process produces heat-depleted liquid stream that must be disposed of in accordance with the appropriate regulations. Most often, this liquid is injected as part of the total reservoir management strategy. In the Imperial Valley, California, high salinity brines are processed by flash crystallizers which produce sludge containing potentially toxic heavy metals such as arsenic, boron, lead, mercury, and vanadium.[39] For example, a 34 megawatt double-flash geothermal power plant tapping the high temperature resource in the Imperial Valley can produce up to 50 tons of sludge every 24 hours.[40] The potential exists for extraction of valuable metals from this sludge prior to disposal, and this option has been explored at some Imperial Valley projects. The DOE has a research and development effort to use bacteria to remove heavy metals from the sludge materials. Some hydrogen sulfide abatement systems produce elemental sulfur which is sold or hauled away at no charge by sulfur producers.

Disposal problems become much more difficult when the waste is toxic. Federal statutes establish land disposal (including reinjection) as the least desirable method of disposal. The Hazardous and Solid Waste Amendments (Public Law 98-616) to the Resource Conservation and Recovery Act (Public Law 94-580) mandate pretreatment of toxic waste to minimize hazards to human health and the environment.

[37]Williams, S., and Porter, K., *Power Plays: Geothermal* (Investor Responsibility Research Center, 1989), p. 178.

[38]Armstead, H.C.H., *Geothermal Energy: Its Past, Present, and Future Contribution to the Energy Needs of Man*, second edition (E.F. Spoon, London, 1983), p. 330.

[39]Armstead, H.C.H., *Geothermal Energy: Its Past, Present, and Future Contribution to the Energy Needs of Man*, second edition (E.F. Spoon, London, 1983).

[40]National Research Council, *Geothermal Energy Technology: Issues, Research and Development Needs, and Cooperative Arrangements* (National Academy Press, Washington, DC, 1987).

Subsidence

Subsidence is hypothesized to occur when large quantities of fluid are withdrawn from reservoirs at liquid-dominated geothermal sites, and the fluid is not reinjected. At some locations, subsidence could become a problem unless sufficiently large quantities of water are injected. In the Imperial Valley of California, valuable farmland might be harmed if subsidence alters local irrigation and drainage patterns.[41] All evidence to date suggests that subsidence is small to nonexistent at U.S. hydrothermal production reservoirs, including those in the Imperial Valley. Along the coast of the Gulf of Mexico, the removal of geopressured fluids might aggravate existing flooding problems. Injection has been successful at preventing subsidence at all liquid-dominated reservoirs. Ongoing geodetic monitoring programs are maintained by developers in tectonically active areas to determine the extent of this potential problem.

Noise Pollution

Noise pollution has been controlled in every instance. At The Geysers, noise pollution became such a problem during well drilling and testing that muffling systems had to be installed.[42] At Steamboat Springs in Nevada, the Yankee-Caithness project limits well testing to business hours to reduce the impact on nearby residents. Noise from power generation equipment is routinely reduced by blanketing and insulating, but complaints are still received concerning generation, pumping, and drilling noise at some sites. Development of resources near population centers may require the type of noise abatement measures used by the oil drilling industry during town-site drilling.

Location Issues

Many of the most promising geothermal resources are located in or near protected areas such as national parks, national monuments, and wilderness, recreation, and scenic areas. The average amount of surface area disturbed for the development of geothermal resources is slight in comparison to other forms of energy extraction. The disturbance usually takes the form of clearcutting of vegetation, grading, and road paving for well pads, pipelines, transmission lines and generation facilities. Erosion and landsliding may be a problem, depending on the steepness of the local terrain.

Environmentalists have tended to oppose all forms of development in protected areas, including geothermal projects. The potential costs of litigation and regulatory compliance associated with the exploitation of these resources will render some exploitation prohibitively expensive.

Geothermal resource development in Hawaii, although technologically promising, has been intensely opposed by some environmental and public interest groups, claiming such development would do irreparable damage to the tropical rain forest while violating local religious beliefs and cultural mores. The emotion-charged controversy could slow the pace of development in Hawaii. Construction of a 27-megawatt plant located there was 65 percent completed at the end of 1990. Generation at this plant is expected to begin in late 1991.

Operational Factors Associated with Hydrothermal Resources

An important area of uncertainty for geothermal developers is the expected operating lifetimes of physical facilities.[43] Since the operating experience of some geothermal technologies is very limited, reservoir life expectancies are an important unknown. The corrosive and scaling potential of geothermal fluids is a significant limiting factor on plant life expectancy. Even though equipment manufacturers try to achieve useful lives of 20 to 30 years, more operating experience is needed to determine if such goals can be attained. Corrosive acids in the geothermal dry steam at The Geysers have destroyed equipment in a matter of weeks.

Geothermal resources can be depleted on a local scale. Several fields, including Wairakei (New Zealand), Larderello (Italy), The Geysers (California), and Heber (California), have experienced slow declines in temperature and pressure over time. However, estimations of the depletion rates or ability of fields to recover are not certain.

[41]Williams, S., and Porter, K., *Power Plays: Geothermal* (Investor Responsibility Research Center, 1989), p. 177.
[42]Williams, S., and Porter, K., *Power Plays: Geothermal*, p. 178.
[43]Williams, S., and Porter, K., *Power Plays: Geothermal*, pp. 165-181.

Significant Milestones Reached in Prediction of Behavior of Injected Fluids

Injection of spent fluids from geothermal power plants is usually necessary to avoid surface discharge of large quantities of fluids, to recharge the reservoir, and to prevent ground subsidence. Effective injection strategies will become even more critical as production declines become more commonplace in mature fields, requiring management of the resource through injection.

Improper injection practices can, however, lead to premature thermal breakthrough in the producing zone, which results in cooling fluid temperature below power plant design limits—an economic disaster. Thus, injection research centers to a large extent on prediction of the behavior of injected fluid through the use of tracers to monitor the migration of the injectate.

The usefulness of the approach has been confirmed by a successful cooperative test by Government researchers and industry. A multiwell, multitracer test was conducted at the Oxbow Geothermal Plant in Dixie Valley, Nevada, by personnel of that company and the University of Utah Research Institute (UURI).

A numerical reservoir model was employed to estimate needed tracer quantities and sampling frequencies as well as to predict test results. Three injection wells were tagged with organic tracers—benzoic acid, benzenesulfonic acid, 4-ethylbenzenesulfonic acid, and florescein—previously tested for this purpose by UURI.

Six production wells were intensively sampled for 2.5 months. During this period, one well showed breakthrough, and the presence of benzoic acid and fluorescein identified the injection well of origin. Concentration ratios of these compounds varied during the test period, as predicted from laboratory experiments. These ratios predicted temperatures consistent with the observed temperatures in the reservoir. Thus, the velocity, direction and effective temperature of the dominant injection-production flowpath in the reservoir were defined.

Other areas of the DOE brine injection technology project involve fluid-rock chemical interactions and injection well placement. The Idaho National Engineering Laboratory (INEL) and Stanford University are applying computer modeling techniques to the tracer return field data for the determination of physical, reservoir properties, and fluid interactions. LBL and UURI are performing theoretical studies of geophysical techniques to determine the effectiveness of existing equipment in detection of theoretically determined signals generated by injected fluids; new equipment will be designed as indicated. INEL is continuing to develop computer models with the capability to analyze and predict the flow of injected fluids and is investigating the potential for coupling the fluid flow computer model with models of chemical interaction between rocks and the injected fluid. As noted elsewhere in this section, UURI is studying potential tracer materials suitable for use in the steam field at The Geysers.

Source: Excerpted from U.S. Department of Energy, *Geothermal Progress Monitor,* No. 12 (December 1990), pp. 15-16.

Market Factors

Electricity Consumption Growth

The EIA projects electricity capacity between 1991 and 1999 in the western United States will increase at a rate of 3.4 percent per year.[44] In this primary marketing region for geothermal power, annual electricity demand growth exceeds the average national growth projected by EIA (approximately 2 percent).[45] Table 8 provides a number of regional electricity demand forecasts from various sources.

Electricity Capacity Growth

Electricity consumption projections are translated into electricity capacity additions by utilities through the development of an integrated resource assessment and through the selection of resource options based on specific need conformance tests established by State regulatory commissions or other regulatory bodies. While this process can be complex, the ultimate goal is to supply sufficient and reliable power at least-cost while ensuring that the utility maintains an adequate return on investment. Currently, the Western Systems

Table 8. Electricity Consumption Forecasts for the Western United States

Forecast Source	Geographical Forecast Region	Forecast Time Horizon	Average Annual Electricity Consumption Growth Rate (Percent)
North American Electric Reliability Council[a]	Western Systems Coordinating Council	1989-1999	1.7
	Northwest Power Pool	1989-1999	1.2
	California/Southern Nevada Area	1989-1999	1.8
	Arizona/New Mexico Area	1989-1999	3.0
California Energy Commission[b]	California	1987-2001	2.2
EIA, *Annual Outlook for U.S. Electric Power*[c]	West Federal Region (CA, NV, AZ)	1989-2000	3.4[d]
	Northwest Federal Region (WA, OR, ID)	1989-2000	2.8[d]
Northwest Power Planning Council[e]	Washington, Oregon, Northern California	1989-2000	-1.0 to 2.7[f]

Sources: Compiled by the Office of Coal, Nuclear, Electric and Alternate Fuels:
[a]North American Electric Reliability Council, *1990 Electricity Supply and Demand* (Princeton, NJ, November 1990).
[b]California Energy Commission, *1990 Electricity Report* (Sacramento, CA, October 1990).
[c]Energy Information Administration, *Annual Outlook for U.S. Electric Power 1990*, Table B9. The base case growth rate is estimated to be about 4.5 percent from 1989 through 1995, about 3 percent from 1996 through 2000, and about 2.4 percent after 2000. Forecast is base case simulation run using the Intermediate Future Forecasting System.
[d]The base case rate is estimated to be about 3.5 percent from 1989 through 1995, about 2.6 percent from 1996 through 2000, and about 2 percent after 2000.
[e]Demand Forecasting Department, Northwest Power Planning Council.
[f]Forecast is provided as a range estimate.

[44]Energy Information Administration, *Electric Power Annual 1989*, DOE/EIA-0348(90) (Washington, DC, January 1991), Table 5, p. 26, Table 10, p. 31.
[45]Energy Information Administration, *Annual Energy Outlook 1991*, DOE/EIA-0383(91) (Washington, DC, 1991), Table A4, p. 48.

Coordinating Council (WSCC) projects that 6,525 megawatts of new installed capacity (at the time of summer peak load) will be needed between 1989 and 2000. This represents a 5.0 percent increase over the 1989 actual installed capacity of 129,533 megawatts.[46]

Energy Supply Competitors

Hydrothermal energy is primarily used to produce baseload electricity and competes with other baseload electricity power production, such as hydropower. Feasibility studies are being conducted to allow hydrothermal electricity to be used in other than base load dispatching modes.

Competition with other energy sources is an important factor for geothermal developers. Large hydroelectric projects in the Canadian province of British Columbia are in the planning stages. Estimates of the cost of Canadian hydroelectric power delivered in the United States run as low as 2 cents per kilowatthour.[47] At that price, Canadian hydroelectric power would be much less expensive than any other power available in the western United States. However, new large-scale hydroelectric projects are proving to be just as difficult to license in Canada as in the United States, and the cost of this new power might be greater than power from indigenous hydrothermal resources. The recent canceling of the British Columbia Hydro Peace River project reduces the likelihood that large amounts of Canadian hydropower will be available to the Pacific Northwest.

Many States are experimenting with competitive bidding systems as a means of awarding power purchase contracts. Under a competitive bidding system, a utility announces a need for power and solicits the delivery of electricity based primarily on price considerations. Other bidding criteria such as dispatchability, power supply reliability, financing, and the likelihood of meeting operational standards, may also be part of the bid evaluation process. These systems may fail to consider the desirable social aspects of renewable resources, such as the favorable environmental impact and reduction of dependence on foreign oil. Competitive bidding favors technologies with both low capital requirements (fossil fuels) and uncertain fuel costs over capital-intensive renewable technologies. Small firms often lack sufficient capital to participate in bidding systems. Despite these potential drawbacks, geothermal power continues to compete successfully for energy supply contracts.

Transmission Access

Access to transmission facilities is a major factor for both utility and nonutility generators.[48] It is a particular problem for some remotely located geothermal producers. Even if access is available, the cost of connecting to the grid system is a highly variable site-specific expenditure. Given this competitive disadvantage, the highest grade and most accessible hydrothermal resources will be exploited first. The extent to which lesser-grade remote resources (the majority of the resource base) will be exploited remains unclear.

Many utilities provide wheeling services to others. However, some prefer not to provide wheeling services on demand for the following reasons: (1) there is inadequate profit incentive to provide wheeling services, and (2) wheeling helps the competition by removing access barriers, thus allowing large users to shop among suppliers. Wheeling is far beyond the capacity of most independent power producers.[49] Hence, they are entirely dependent on the indulgence of the closest utilities for such services. Lacking sufficient incentive, utilities have in a few cases refused to enter into wheeling transactions, dampening the optimism of other developers.

Factors Associated with the Development of Geopressured Resources, Hot Dry Rock, and Magma

This section briefly discusses some factors which could affect the development of other types of geothermal resources. The direct extraction of energy from magma has been the subject of research for many years.[50] While a single volcano contains a huge concentration of energy within a relatively small geographical area, formidable technical problems prevent the exploitation of magma resources.

[46]North American Electricity Reliability Council, *1990 Electric Supply and Demand* (Princeton, NJ, November 1990), p. 18.

[47]"Questions Remain on the Expansion of the Pacific Northwest-Southwest Intertie," *Energy*, Winter 1989, vol. 24, pp. 25-27.

[48]McCullough, R., "Establishing the Electric Pipeline: The Role of Energy Brokers," *Public Utilities Fortnightly*, December 6, 1990, vol. 126, pp. 34-37.

[49]McCullough, R., "Establishing the Electric Pipeline: The Role of Energy Brokers."

[50]Armstead, H.C.H., *Geothermal Energy: Its Past, Present, and Future Contribution to the Energy Needs of Man*, second edition (E.F. Spoon, London, 1983), p. 361.

"Consensus" Legislation Enacted to Establish Newberry National Volcanic Monument

In order to protect the spectacular natural features of the Newberry Caldera in central Oregon and at the same time permit geothermal development in adjacent areas, a local committee hammered out the provisions of "consensus" Federal legislation to establish the Newberry National Volcanic Monument which became law on November 5, 1990 (Public Law 101-522). This effort took "two laborious years," according to witnesses before the House Subcommittee on National Parks and Public Lands on June 18, 1990, an effort which was rewarded with prompt action in both the House and Senate.

Through the efforts of the U.S. Forest Service and many others, the National Monument Committee was formed, consisting of about 30 people representing environmental interests, users such as snowmobilers and hunters, and commercial interests such as timber, tourism, and geothermal energy as well as Federal, State, and local governments. The Committee and its various subcommittees held hundreds of meetings to arrive at what is called a "win-win" situation.

The legislation places a large block of acreage in the Deschutes National Forest in the Cascades Range into the national monument. This acreage embraces parts of the Newberry Caldera Known Geothermal Resources Area (KGRA), designated in 1974, which is considered to be the prime geothermal prospect in the Pacific Northwest. It is now withdrawn from use of mining or disposal under all mineral and geothermal leasing laws.

Since 1982, several geothermal leases were issued for the flanks of the volcano outside the KGRA, but no leases have been issued within the KGRA. The Forest Service had closed the acreage within the crater rim to development, and leases on the remaining acres were awaiting completion of Environmental Impact Statements to determine where and under what conditions leasing should occur. The compromise legislation cancels existing geothermal leases within the monument and directs the Secretary of the Interior to issue leases on other lands as full compensation.

Use of the acreage immediately surrounding the monument for commercial acts recognizes the importance of the underlying geothermal resource in making special provisions for its exploitation. A "Transferral Area" is withdrawn from further leasing until completion of a well capable of producing geothermal steam in commercial quantities, as defined by the Geothermal Steam Act, on a valid existing lease; at that time the withdrawal would be revoked. Use of areas designated as "Transferral Area Adjacent" and "Special Management Area" are similarly dependent upon actual discovery of a commercial resource, and leases therein would carry a "No Surface Occupancy" stipulation requiring directional drilling from other leases. The Secretary is directed to hold a competitive lease sale for lands within the "Geothermal Lease Sale Parcels" within 1 year.

The language of the act establishing these various areas is tied to a map of the area, which can be seen at the office of the Deschutes National Forest in Bend, Oregon (Sally Collins, 503/388-2715), or at Forest Service Headquarters in Washington (Gene Zimmerman, 202/382-8215).

The geothermal industry will forego a major percentage of the potential resource value in the area, and the Federal Government will potentially lose millions of dollars in royalties, but both parties express satisfaction with the legislation.

Source: U.S. Department of Energy, *Geothermal Progress Monitor*, No. 12 (December 1990), p. 41.

The very high temperatures encountered adjacent to magma bodies can cause drilling equipment to fail. The reaction of dissolved gases to a sudden release of pressure by the drill hole can be explosive. Even if some method of penetrating the rock immediately adjacent to the magma body is found, a heat extraction technology must be developed. The underlying assumption is that the great intensity of heat within magma bodies will yield sufficient quantities of energy to justify the anticipated high cost of extraction.[51] Commercial development of magma resources remains in the distant future. For this reason, the DOE has deferred further research into magma energy extraction technology in order to concentrate on near-term needs.

The extraction of energy from geopressured geothermal resources has also been the subject of research that culminated in the construction of one small demonstration plant near Pleasant Bayou, Texas. The plant was operated for 1 year and is no longer in operation. To support a commercially viable enterprise, the temperature of the fluid must be sufficiently high, and there must be a sufficient quantity of dissolved methane. The reservoir must be sufficiently large and permeable to allow adequate production of fluids over an extended period of time. These issues have led to reasoned speculation that only a limited portion of the U.S. geopressured resource may be economically exploitable for the foreseeable future.

Hot dry rock technology has progressed beyond the feasibility stage and presently is in the demonstration phase. Research has shown that the resource can be reached at economic depths; that hydraulic fracturing can be effectively used to create man-made reservoirs in hard rock; and that heat can be extracted from such reservoirs utilizing water as a working fluid. However, the geology of hot dry rock resource areas varies, and the technology to develop man-made reservoirs in different geologic conditions is unproven and might be expensive. Although hot dry rock resources have the potential to yield enormous quantities of energy, the path to exploitation requires significant technical developments. It may be 20 years before significant quantities of electricity can be produced from hot dry rocks, depending on the level of effort to resolve outstanding issues.

The term "heat mining" was coined[52] to describe the process of extracting heat energy from hot dry rock. Three requirements must be satisfied before "heat mining" will be commercially viable: (1) the development of inexpensive high-temperature hard-rock drilling techniques, (2) improvements in three-dimensional rock fracturing, and (3) mastery of methods of maintaining low-impedance fluid circulation through the fracture system. Efforts to satisfy these requirements are being made by the Department of Energy.

A major technical obstacle to exploitation is the development of a method for extracting heat from deeply buried rock. The hot dry rock resource base occurs in igneous and metamorphic terrains containing rocks that lack sufficient matrix or fracture permeability for the migration of fluids. Under those circumstances it is necessary to create an extensive interconnected fracture system which allows: (1) sufficient fluid circulation, (2) fluid removal, and (3) fluid reinjection. Recent tests of hydraulically created fracture systems have successfully created adequate circulation. After the permeable zone has been created, water must be injected into the formation. The quantity of water needed is uncertain, nor is it known to what extent circulating fluids will precipitate scale in fracture systems.

[51]National Research Council, *Geothermal Energy Technology: Issues, Research and Development Needs, and Cooperative Arrangements* (National Academy Press, Washington, DC, 1987).

[52]Armstead, H.C.H., *Geothermal Energy: Its Past, Present, and Future Contribution to the Energy Needs of Man*, second edition (E.F. Spoon, London, 1983), p. 348.

The DOE Geothermal Exploration R&D Program

The Department of Energy has supported many important and exciting R&D programs over the past several years, relating to drilling and production of geothermal fluids, fluid treatment, energy conversion, and future energy sources. All of this is as it should be. Exploration is encompassed in certain aspects of the research into magma energy and hot dry rock; and exploration is assessed inferentially in the course of well logging or drilling technology. There is, however, almost no geothermal exploration program *per se*. The reason for this is not hard to see: there is almost no demand for exploration of new geothermal systems in the United States today.

During the 1960's and 1970's, exploration companies were remarkably successful in discovering geothermal fields in a variety of geologic settings in the western United States. However, the market for geothermal power lagged far behind the discovery rate. With the decline in energy prices in the 1980's, exploration all but ceased. All emphasis was placed upon the commercial development of already discovered fields; and there were very many of them, as the result of American technical ingenuity and persistence. This trend to commercialization was enhanced by various government policies ((PURPA) Standard Offer contracts, for example), and by the introduction of binary-cycle technology to allow utilization of low-temperature systems (120-180°C) that previously had been dismissed as uneconomical.

The 1980's, therefore, was the decade of commercial development. However, an interesting thing now has occurred: almost all of the previously discovered, easily accessible geothermal fields in the United States have been committed for development. There are still a few discovered fields waiting for commercialization in the Cascade Range of California and Oregon, and at a couple of places in Nevada and New Mexico. Once these have been contracted for development, there is just the fill-in drilling and development to be done at other fields. That is, unless there is a new program of exploration by industry and government.

The exploration successes of the 1960's and 1970's were based on a very simple formulation: start where boiling water or steam is pouring out of the ground, and explore and drill outward from those spots. Simple and effective; except in those places where cold groundwater masked the upward flow of hot fluid, or where steep topography resulted in long outflow tongues, far from the geothermal upwelling, or where complex mixing of several non-thermal and thermal aquifers was occurring. For the 1990's, we will still have a few undrilled areas where steam or boiling water comes out of the ground at us. But we will need to accept the challenge that other geothermal systems may have much more subtle signatures, and perhaps cannot be found by our traditional techniques.

The astonishing thing in a brief retrospective view of geothermal exploration is how few fields were discovered by geophysics, and how many were discovered by almost-random drilling of water wells, temperature-gradient holes, oil tests, and geothermal go-for-broke exploration wells. Perhaps this was inevitable in an industry that had no antecedent methodology, but which borrowed from the oil industry, and improvised as it went along.

We see today that temperature-gradient drilling, to ever greater depth, has become the standard method for exploration in the Cascade Range, the Basin and Range, even The Geysers, and now in Hawaii. However, drilling, even drilling slim holes, is expensive; and with larger target circles being drawn, now that the boiling springs are largely drilled, drilling may not be sufficiently regional a tool for the future.

Source: J.B. Koenig, GeothermEx, Inc. (1990).

Appendix A
Geothermal Resources

Geothermal energy is the heat of the earth. Geothermal resources which might be used to generate electricity result from complex geologic processes that lead to heat being concentrated at accessible depths. Hydrothermal energy,[53] geopressured energy, and magma energy, all result from this concentration of earth's heat in discrete regions of the subsurface. Temperature within the earth increases with increasing depth. Highly viscous or partially molten rock[54] at temperatures between 1,200°F and 2,200°F (650°C and 1,200°C) is postulated to exist everywhere beneath the earth's surface at depths of 50 to 60 miles (80 to 100 kilometers), and the temperature at the earth's center, nearly 4,000 miles (6,400 km) deep, is estimated to be 7,200°F (4,000°C) or higher. Because the earth is hot inside, heat flows steadily outward and is permanently lost from the surface by radiation into space.

Earth energy is thermal energy at the normal temperature of the shallow ground and is found everywhere across the United States. Earth energy can be used with geothermal heat pumps. The mean value of surface heat flow is 0.082 watts[55] per square meter (W/m^2), commonly stated in milliwatts per square meter (mW/m^2) as 82 mW/m^2. Because the surface area of the earth is 5.1×10^{14} m^2, the rate of heat loss is about 42 million megawatts.[56] The heat flux from the earth's interior is about 5,000 times smaller than the radiation we receive from the sun (much of which is reflected, or radiated back into space). Thus, the earth's surface temperature is controlled by the amount of heat we receive and retain from the sun and not by internal heat.[57] However, within a few meters below the surface the sun's influence disappears.

Three sources of internal heat are most important: (1) heat released from decay of naturally radioactive elements throughout the earth's 4.7-billion-year history; (2) heat of impact and compression released during the original formation of the earth by accretion of in-falling meteorites; and (3) heat released from the sinking of abundant heavy metals (iron, nickel, copper) as they descended to form the earth's core. An estimated 45 to 85 percent of the heat escaping from the earth is due to radioactive decay of elements concentrated in the crust.[58,59] The remainder is due to slow cooling of the earth, with heat being brought up from the core by convection in the viscous mantle.[60]

A schematic cross section of the earth is shown in Figure A1. A solid layer, the lithosphere,[61] extends from the surface to a depth of about 100 km (62 miles). The lithosphere is composed of an upper layer, called the crust, and the uppermost regions of the mantle, the unit that lies below the crust. The lithosphere is solid rock, but the mantle material below the lithosphere behaves as a very viscous liquid due to its high temperature and pressure. It will flow very slowly under sustained stress. The outer core is believed to be composed of a liquid iron-nickel-copper mixture while the inner core is a solid mixture of these metals.

Geological Processes

The genesis of geothermal resources lies in the geological transport of anomalous amounts of heat close enough to the surface for access. Thus, the distribution of geothermal areas is not random but is

[53]Definitions for this and other technical terms can be found in the Glossary.

[54]*Viscous rock* is rock which flows in an imperfectly fluid manner upon application of unbalanced forces. The rock will change its form under the influence of a deforming force, but not instantly, as more perfect fluids appear to do.

[55]A *watt (thermal)* is a unit of power in the metric system, expressed in terms of energy per second (see Glossary).

[56]Williams, D.L., and Von Herzen, R.P., "Heat Loss from the Earth—New Estimate," *Geology*, vol. 2, 1974, pp. 327-328.

[57]Bott, M.H.P., *The Interior of the Earth—Its Structure, Constitution and Evolution* (London: Edward Arnold, 1982), 403 pp.

[58]The *crust (crustal zones)* is the outer layer of the earth, originally considered to overlay a molten interior, now defined in various ways (lithosphere, tectonosphere, etc.).

[59]Bott, M.H.P., *The Interior of the Earth—Its Structure, Constitution and Evolution* (London: Edward Arnold, 1982), 403 pp.

[60]The *mantle* is the layer of the earth lying between the crust and the core. The mantle extends between depths of about 19 miles (30 km) in the continental areas and 1,790 miles (2,800 km), where the core begins.

[61]The *lithosphere* is the upper, solid part of the earth. It includes the crust and uppermost mantle.

Figure A1. Schematic of the Earth's Interior

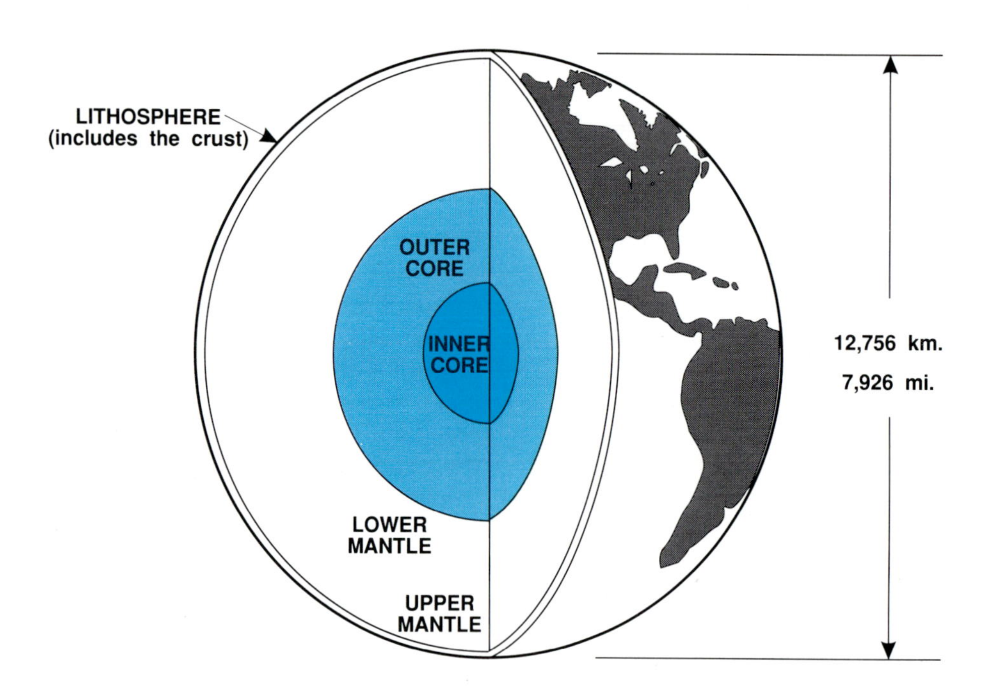

Source: Modified from Armstead, H.C.H., *Geothermal Energy: Its Past, Present, and Future Contribution to the Energy Needs of Man*, second edition (E.F. Spoon, London, 1983), p. 21.

Figure A2. Schematic of Tectonic Plate Movements

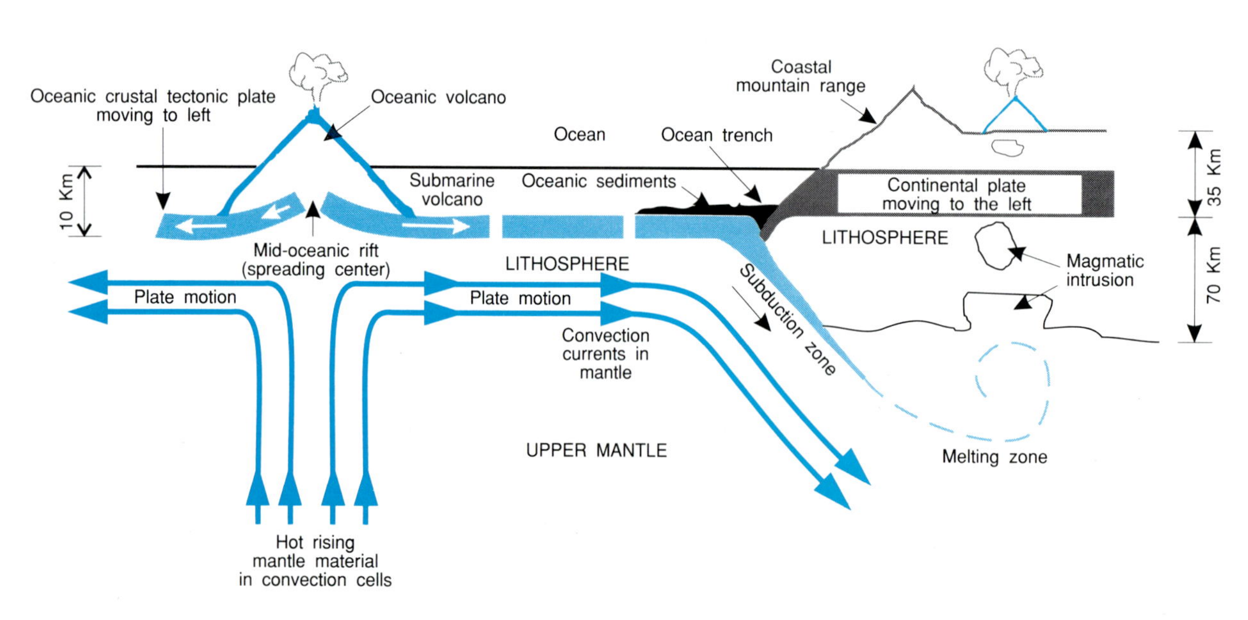

Note: Not drawn to scale.
Source: Modified from Petroleum Information Corporation, *The Geothermal Resource* (A.C. Nielsen Co., 1979).

governed by geological processes of global, regional and local scale. This fact is important in exploration for geothermal resources.

Geothermal resources commonly have three components: (1) an anomalous concentration of heat (i.e., a *heat source*); (2) *fluid* to transport the heat from the rock to the surface; and (3) *permeability* in the rock sufficient to form a plumbing system through which the water can circulate.

Heat Sources

The two most common sources of geothermal heat are: (1) intrusion of molten rock (magma) from great depth to high levels in the earth's crust; and (2) ascent of groundwater that has circulated to depths of 1 to 3 miles (1.6 to 5 km) and has been heated in the normal or enhanced geothermal gradient without occurrence of a nearby intrusion.

One geological process that generates shallow magmatic crustal intrusions in several different ways is known as plate tectonics[62] (Figure A2). Outward heat flux from the deep interior forms convection cells in the mantle in which hotter material, being less viscous and more buoyant than surrounding material, slowly rises, spreads out under the solid lithosphere, cools and descends again. The lithosphere cracks above the areas of upwelling and is split apart along linear or arcuate structures[63] called *spreading centers,*[64] which occur for the most part in the ocean basins. Due to this mechanism, the earth's lithosphere is broken into about 12 large, rigid plates. The *spreading plate boundaries* are zones typically thousands of miles long and several hundred miles wide characterized by major rifts or faults and coincident with the world's mid-oceanic mountain and rift system. Crustal plates on each side of the central rift zone separate at a rate of a few centimeters per year, and molten mantle material rises in the crack, where it solidifies to form new oceanic crust. Seismic activity[65] from southern California to Alaska is a direct consequence of this plate motion.

Since new crust is being created at spreading centers while the circumference of the earth remains constant, crust must be consumed somewhere. As the laterally moving oceanic plates press against neighboring plates, some of which contain the imbedded continental land masses, the oceanic plates are thrust beneath the continental plates. These zones of under-thrusting, where crust is consumed, are called *subduction zones.*[66] They are marked by the world's deep ocean trenches, formed as the sea floor is dragged down by the subducted oceanic plate.

The subducted plate descends into the mantle and is heated by the surrounding warmer material and by friction. Temperatures become high enough to cause partial melting. Since molten or partially molten rock bodies (magmas) are lighter than solid rock, the magmas ascend buoyantly through the crust. Volcanos result if some of the molten material escapes at the surface, but the majority of the magma usually cools and consolidates underground. Since the subducted plate descends at an angle of about 45 degrees, crustal intrusion and volcanos occur on the landward side of oceanic trenches 30 to 150 miles (50 to 250 km) inland. The volcanos of the Cascade Range of California, Oregon, and Washington, for example, overlay the subducting Juan de Fuca plate and owe their origin to the process just described. The Pacific Ring of Fire, which extends around the margins of the Pacific basin, is composed of volcanos in the Aleutians, Japan, the Philippines, Indonesia, New Zealand, South America, and Central America, all of which are due to subduction.

Oceanic rises, where new crustal material is formed, occur in all major oceans. The East Pacific Rise, the Mid-Atlantic Ridge and the Indian Ridges are examples. In places, the ridge crest is offset by large *transform faults*[67] that result from variations in the rate of spreading along the ridge. Figure A3 is a conceptual map showing the distribution of tectonic regions in North America and the eastern Pacific Ocean, where most of the observed geothermal activity of the continent can be found.

[62]*Plate tectonics* is a theory of global-scale dynamics involving the movement of many rigid plates of the earth's crust. Considerable tectonic activity occurs along the margins of the plates, where buckling and grinding occur as the plates are propelled by the forces of deep-seated mantle convection currents. This has resulted in continental drift and changes in the shape and size of oceanic basins and continents.

[63]*Arcuate structures* are geologic formations that are curved or bowed.

[64]*Spreading centers* are the cracks in the lithosphere over extended distances above areas of upwelling.

[65]*Seismic activity* is the phenomenon of earth movements (earthquakes).

[66]Subduction zones are elongate regions along which a crustal block descends relative to another crustal block.

[67]A *transform fault* is a strike-slip fault characteristic of mid-oceanic ridges and along which the ridges are offset.

Figure A3. Tectonic Regions in North America and the Eastern Pacific Ocean

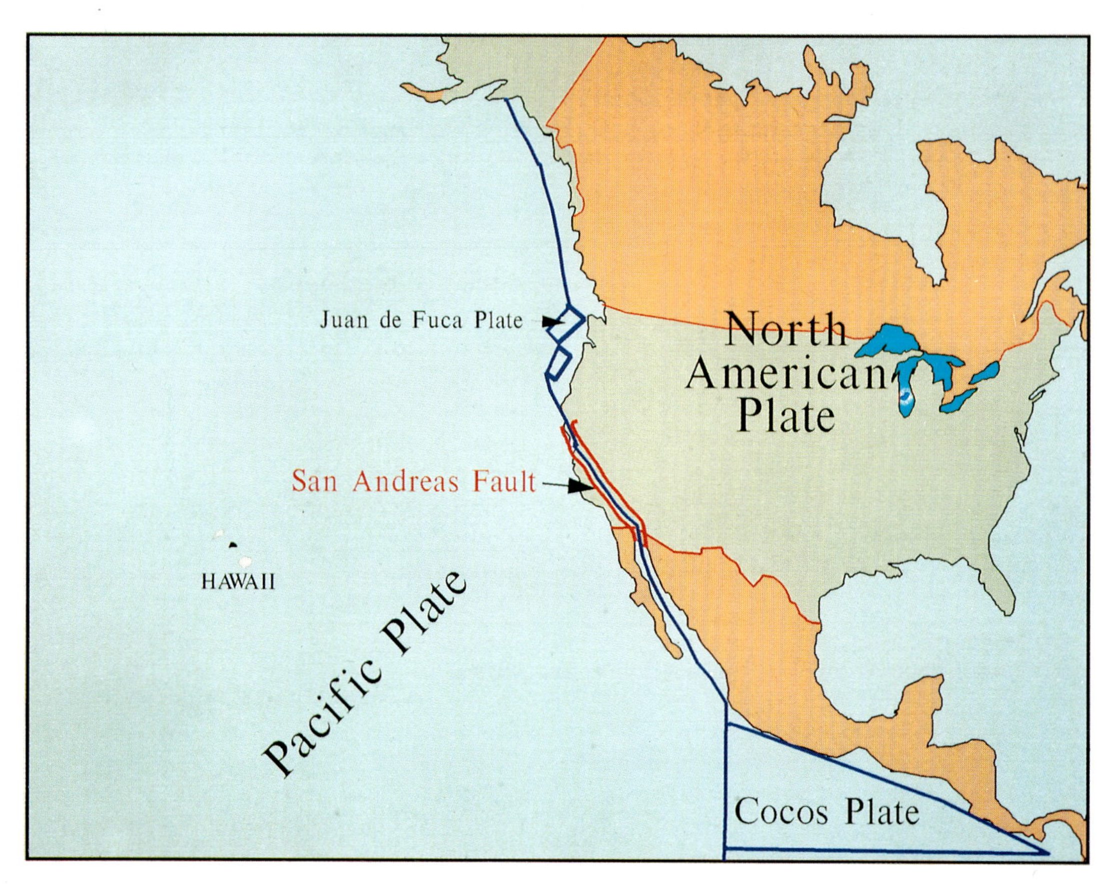

Sources of information used in developing the map: Armstead, H.C.H., *Geothermal Energy: Its Past, Present, and Future Contribution to Energy Needs of Man*, second edition (E.F. Spoon, London, 1983); Leet, L.D., Judson, S., and Kauffman, M., *Physical Geology*, 5th edition (Prentice Hall, 1978); Rybach, L., "Geothermal Systems, Conductive Heat Flow, Geothermal Anomalies," in *Geothermal Systems: Principles and Case Histories*, eds., L. Rybach and L.J.P. Muller (Wiley Publishing, NY, 1981), pp. 3-36; and Simkin, T., Tilling, R., Taggart, J., Jones, W., and Spall, H., *This Dynamic Planet: World Map of Volvanoes, Earthquakes, and Plate Tectonics* (Smithsonian Institution and U.S. Geological Survey, Washington, DC, 1989).

There is a close correlation between both spreading and subducting plate boundaries and the locations of geothermal occurrences. Both spreading and subduction result in earthquakes. Seismicity is believed to be helpful in keeping faults and fractures open in the rocks for geothermal waters to circulate.

Another important source of volcanic rocks are point sources of heat in the mantle. The mantle contains local areas of upwelling, hot material called *plumes*,[68] which have persisted for millions of years. As crustal plates move over these hot spots, a linear or arcuate chain of volcanos results, with young volcanic rocks at one end of the chain and older ones at the other end. The Hawaiian Island chain is an example. The thermal features of Yellowstone National Park are believed to be the result of an underlying mantle plume.

Fluid

Geothermal resources require a *fluid transport medium*.[69] In the earth that medium is groundwater that circulates near or through the heat source. The groundwater can originate as connate water[70] that was trapped in voids during the formation of the rock. But quite often the water is meteoric in origin, meaning it percolated from the surface along pathways determined by geological structures such as faults and formation boundaries.

[68]A *plume* is a body of magma that upwells in localized areas.

[69]A *fluid transport medium* is a liquid that transports energy, dissolved solids, or dissolved gases from their origin to their destination.

[70]*Connate water* is water entrapped in sedimentary rock at the time the rock was deposited. It may be derived from either ocean or land water.

The density and viscosity of water both decrease as temperature increases. Water heated at depth is lighter than cold water in surrounding rocks, and is therefore subject to buoyant forces that tend to push it upward. If heating is great enough for buoyancy to overcome the resistance to flow in the rock, heated water will rise toward the earth's surface. As it rises, cooler water moves in to replace it. In this way, natural convection is set up in the groundwater around and above a heat source such as an intrusion. Convection can bring large quantities of heat within reach of wells drilled form the surface.

Because of their varied origin and the reactivity inherent to heated water, geothermal waters exhibit a wide range of chemical compositions. Salinities can range from a few parts per million up to 30 percent; dissolved gases such as carbon dioxide and hydrogen sulfide are common. As a result, geothermal waters play an important role in crustal processes, not only in transporting heat, but also in altering the physicochemical properties of rock. Such fluids have produced many ore deposits of copper, lead, zinc, and other metals in proximity to heat sources.

Permeability

Permeability is a measure of a rock's *capacity to transmit fluid*. The flow takes place in pores between mineral grains and in open spaces created by fractures and faults. *Porosity* is the term given to the *amount of void space* in a volume of rock. Interconnected porosity provides flowpaths for the fluids, and creates permeability. In addition to the pore spaces, structures that form porosity and permeability include open fault zones, fractures and fracture intersections, contacts between different rock types and shattered zones produced by hydraulic fracturing, and mineral growth areas in rocks.

Most geothermal systems are structurally controlled, i.e., the magmatic heat source has been emplaced along zones of structural weakness in the crust. Permeability may be increased around the intrusion from fracturing and faulting in response to stresses involved in the intrusion process itself and in response to regional stresses.

Classification of Geothermal Resources

Geothermal resources are usually classified as shown in Table A1. A cross-section of the earth showing a typical source of geothermal energy is provided in Figure A4. Geothermal resources are divided into the following ranges: *low* temperatures (<90°C or 194°F), *moderate or intermediate* temperatures (90°C to 150°C or 194°F to 302°F), and *high* temperatures (>150°C or 302°F).

Convective Hydrothermal Resources

Convective hydrothermal resources are geothermal resources in which the earth's heat is carried upward by convective circulation of naturally occurring hot water or steam. Underlying some localized high-temperature convective hydrothermal resources (Table A2) is presumably an intrusion of still-molten or recently solidified rock at a temperature between 1,100°F and 2,000°F (600°C and 1,100°C). Other convective resources result from circulation of water

Table A1. Classification of Geothermal Resources

Resource Type	Temperature Characteristics
Convective Hydrothermal Resources	
Vapor-Dominated	About 240°C (about 460°F)
Hot-Water-Dominated	20 to 350+°C (70 to 660°F)
Other Hydrothermal Resources	
Sedimentary Basin / Regional Aquifers (hot fluid in sedimentary rocks)	20 to 150°C (70 to 300°F)
Geopressured (hot fluid under high pressure)	90 to 200°C (190 to 400°F)
Hot Rock Resources	
Part Still Molten (magma)	>600°C (>1,100°F)
Solidified	90 to 650°C (190 to 1,200°F)

Source: Modified from White, D.W., and Williams, D.L., *Assessment of Geothermal Resources of the United States—1975*, U.S. Geological Survey Circular 726 (Washington, DC, 1975), pp. 147-155.

Figure A4. Cross-section of the Earth Showing Source of Geothermal Energy

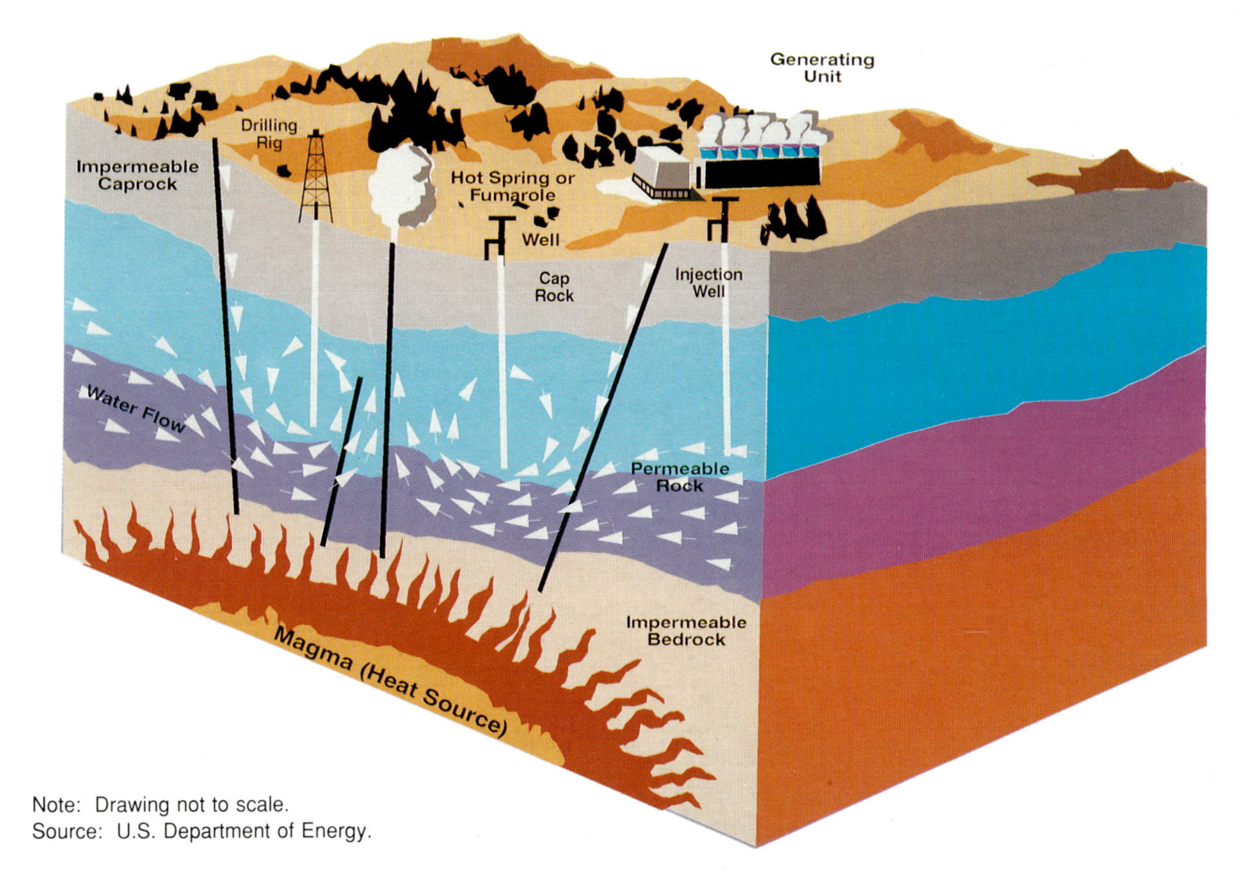

Note: Drawing not to scale.
Source: U.S. Department of Energy.

down fractures to depths where the rock temperature is elevated even in the absence of an intrusion, with heating and buoyant transport of the water to the surface.

A conceptual model of a hydrothermal system where *steam* is the pressure-controlling phase is a so-called *vapor-dominated geothermal system.*[71] The Geysers geothermal area in California, about 80 miles north of San Francisco, is a vapor-dominated resource. Steam is produced from depths of 3,000 to 10,000 feet (1 km to 3 km) and is used to run turbine engines which turn electrical generators. The Geysers is the largest geothermal electrical producing area in the world, with a capacity exceeding 2,000 megawatts of power. Other producing vapor-dominated resources occur at Larderello and Monte Amiata, Italy, and at Matsukawa, Japan.

In a high-temperature, *liquid-dominated geothermal system,*[72,73,74,75] groundwater circulates downward in open fractures and removes heat from deep, hot rocks as it rises buoyantly and is replaced by cool recharge water moving in from the sides. Rapid convection produces uniform temperatures over large volumes of the reservoir. There is typically an upflow zone at the center of each convection cell, an outflow zone or plume of heated water moving laterally away from the center of the system, and a downflow zone where recharge water is actively moving downward.

[71]White, D.E., Muffler, L.J.P., and Truesdell, A.H., "Vapor-Dominated Hydrothermal Systems Compared with Hot-Water Systems," *Economic Geology*, vol. 66, 1971, pp. 75-97.

[72]White, D.E., and Williams, D.L., eds., *Assessment of Geothermal Resources of the United States—1975*, U.S. Geological Survey Circular 726, 1975, 155 pp.

[73]Mahon, W.A.J., Klyen, L.E., and Rhode, M., "Neutral Sodium/Bicarbonate/Sulfate Hot Waters in Geothermal Systems," *Chinetsu* (Journal of the Japan Geothermal Energy Association), vol. 17, 1980, pp. 11-24.

[74]Henley, R.W., and Ellis, A.J., "Geothermal Systems Ancient and Modern, a Geochemical Review," *Earth Science Review*, vol. 19, 1983, pp. 1-50.

[75]Norton, D.L., "Theory of Hydrothermal Systems," *Annual Reviews of Earth and Planetary Science*, vol. 12, 1983, pp. 155-177.

Table A2. Estimated Physical Characteristics for Selected Identified High-Temperature Hydrothermal Resources

Resource Area	Mean Temperature[a] (degrees C)	Mean Reservoir Area[b] (square kilometers)
Salton Sea, California	323	60
Mono-Long Valley, California	227	82
The Geysers, California	237	100
Roosevelt, Utah	265	24
Coso Hot Springs, California	220	27
East Mesa, California	182	33

[a]Muffler, L.J.P., editor, *Assessment of Geothermal Resources of the United States—1978*, U.S. Geological Survey Circular 790 (1979), pp. 44-57.

[b]Muffler, L.J.P., editor, *Assessment of Geothermal Resources of the United States—1978*, U.S. Geological Survey Circular 790 (1979), p. 30.

Escape of hot fluids is often minimized by a near-surface sealed zone or caprock formed by precipitation of minerals in fractures and pore spaces.

The geothermal reservoir is the volume containing hydrothermal fluids at a useful temperature. The porosity of the reservoir rocks determines the total amount of fluid available, whereas the permeability determines the rate at which fluid can be produced. One must *not* envisage a large bathtub of hot water that can be tapped at any handy location. Both porosity and permeability vary over wide ranges at different points in the reservoir. A typical well will encounter impermeable rocks over much of its length, with steam or hot water inflow mainly along a few open fractures or over a restricted rock interval. Apertures of producing fractures are sometimes as small as 1/16 inch, but in other cases they reach 1 inch or more. Areas where different fracture or fault sets intersect or where fractures intersect permeable rock units may be especially favorable for production of large volumes of fluid. The longevity of a well depends upon how completely the producing zones are connected to the local and reservoir-wide network of porosity.

Geopressured Resources

Geopressured resources occur in basin environments. They consist of deeply buried fluids contained in permeable sedimentary rocks warmed in a normal or enhanced geothermal gradient. The fluids are tightly confined by surrounding impermeable rock, where the fluid pressure supports a portion of the weight of the overlying rock column as well as the weight of the water column. A large resource of geopressured fluids occurs along the Gulf Coast of the United States, where the geopressured waters generally contain dissolved methane. Three sources of energy are actually available from these resources: heat, mechanical energy caused by the great pressure with which these waters exit the borehole, and recoverable methane.

The major obstacle to the economic development of geopressured geothermal resources is the high cost of drilling deep, high-pressure wells or converting wells drilled originally for oil and gas exploration. The Electric Power Research Institute and the DOE jointly funded a 1-megawatt geopressured demonstration plant in Texas that generated electricity with an internal combustion engine fueled by extracted methane. At the same time, heat from the geothermal brine was combined with exhaust gas in the binary cycle to generate additional power. This hybrid power system showed conversion efficiency improvements of 15 to 20 percent over standard binary or gas turbine systems.

Hot Dry Rock Resources

Hot dry rock resources are defined as heat stored in rocks within about 6 miles (10 km) of the surface from which the energy cannot be economically extracted by natural hot water or steam. These hot rocks have few pore spaces or fractures, and therefore, contain little water and little or no interconnected permeability. The feasibility and economics of extraction of heat from hot dry rock has, for more than a decade, been the subject of a research program at the Department of Energy's Los Alamos National Laboratory in New Mexico.[76] An experimental site has been established at Fenton Hill, on the edge of the Valles Caldera in New Mexico. Similar research has been done in England.[77] Both projects indicate that it is technologically feasible to induce a permeable fracture system in hot impermeable

[76]Hendron, R.H., "The U.S. Hot Dry Rock Project," in *Proceedings of the Twelfth Workshop on Geothermal Reservoir Engineering* (Stanford University, 1987), pp. 7-12.

[77]Batchelor, A.S., "The Stimulation of a Hot Dry Rock Geothermal Reservoir in the Cornubian Granite, England," in *Proceedings of the Eighth Workshop on Geothermal Reservoir Engineering*, SGP-TR-60 (Stanford University, 1982), pp. 237-248.

rocks through hydraulic fracturing from a deep well. During formation of the fracture system, its dimensions, location and orientation are mapped using geophysical techniques. A second borehole is located and drilled such that it intersects the hydraulic fracture system. Water can then be circulated down one hole, through the fracture system where it removes heat from the rocks, and up the second hole (Figure A5).

The principal aim of the research at Los Alamos is to develop the engineering data needed to evaluate the economic viability of candidate resources. The current plans are for a long-term flow test of the existing two-well system in order to determine production characteristics of the artificially created fracture system and its thermal drawdown and rate of water loss.

Figure A5. Hot Dry Rock (HDR) Geothermal System Concept for Low-Permeable Formations

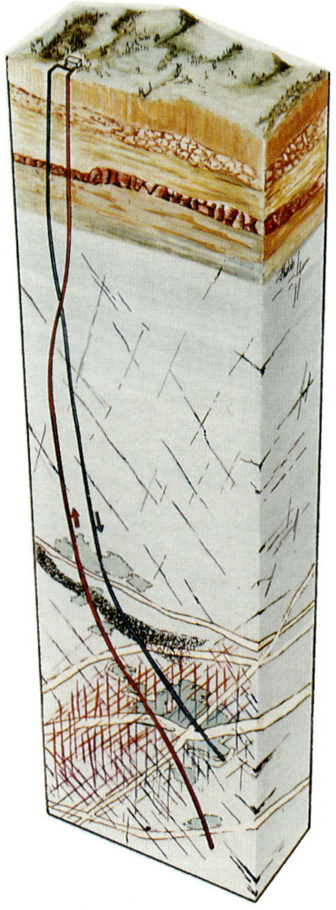

Surface plant

Cold water (25°C) circulated down to hot rock reservoir

Directionally drilled wells to intercept fracture zone

Hydraulic fracture zone (10,000 square feet)

Hot water (200°C) produced from second well

10,000 feet

Reservoir, 225 to 300°C (depends on depth and location)

Note: Drawing not to scale.
Source: U.S. Department of Energy.

Molten Rock (Magma) Resources

The extremely high temperature of magma makes it attractive as a geothermal resource. Even though there is no existing commercial technology that allows heat recovery from these resources, the potential exists for large amounts of electrical energy production from a single well.[78]

Distribution of Hydrothermal Resources in the United States

Figure A6 displays selected commercial geothermal resources in the Western United States. Heat flow contours for the Western United States (Figure A7) give a rough approximation of where unidentified geothermal resources might exist.[79,80] There are many more low- and moderate-temperature resources than high-temperature resource.[81] Not shown are locations of hot dry rock or magma resources because the resources have yet to be identified systematically. In fact, the present knowledge of the geothermal resource base for all types of geothermal resources, except earth heat, is poor. Earth heat resources, used as the source and/or sink for the operation of geothermal heat pumps, occur everywhere.

The DOE maintains an active, broad-based R&D program aimed at developing the technology needed to extract and use hydrothermal resources in an economic and environmentally benign manner. This program includes research in exploration methods, reservoir technology, drilling, energy conversion, materials, chemistry, and waste management. Since most of the easily located geothermal systems are already known and many of those are developed, a new generation of exploration technologies are being developed to locate and characterize the hidden geothermal systems which do not reach the surface. In addition to those new geophysical and geochemical methods, new procedures

[78]Dunn, J.C., Ortega, A., Hickox, C.E., Chu, T.Y., and Wemple, R.P., "Magma Energy Extraction," in *Proceedings of the Twelfth Workshop on Geothermal Reservoir Engineering* (Stanford University, 1987), pp. 13-20.

[79]Muffler, L.J.P., editor, *Assessment of Geothermal Resources of the United States—1978*, U.S. Geological Survey Circular 790, 1979, 163 pp.

[80]Reed, M.J., editor, *Assessment of Low-Temperature Geothermal Resources of the United States—1982*, U.S. Geological Survey Circular 892, 1983, 73 pp.

[81]Reed, M.J., editor, *Assessment of Low-Temperature Geothermal Resources of the United States—1982.*

Figure A6. Selected Geothermal Resource Areas in the Western United States

Source: Modified from Petroleum Information Corporation, *The Geothermal Resource* (A.C. Nielsen Co., 1979), p. 9.

for reservoir testing and evaluation compatible with core drilling, are being examined. Improved methods will also require new means of data interpretation. Significant growth in geothermal development will rely on the discovery and production of several new water-dominated geothermal fields.

Resource Assessments

The assessment of the quantity, quality, and distribution of domestic geothermal resources is basic to the analysis of geothermal energy's potential role in electric power generation. The varying physical characteristics described above contribute to different qualities of geothermal resources. As expected from their record of commercial exploitation, hydrothermal resources have been assessed in more detail than other types of geothermal resources.

The process of finding, delineating, and developing geothermal resources is outlined in Figure A8. Resource estimates are based on analyses of indicators from geology, geothermometry, geochemistry, geophysics, and downhole measurements. Many geothermal fields have not been adequately drilled to delineate their contained resources with reasonable certainty. Where numerous well data and attendant feasibility studies are available for hydrothermal fields, reserve estimates have been developed with the degree of confidence necessary to acquire financing for development projects.

Figure A7. Heat Flow Contours for the Western United States

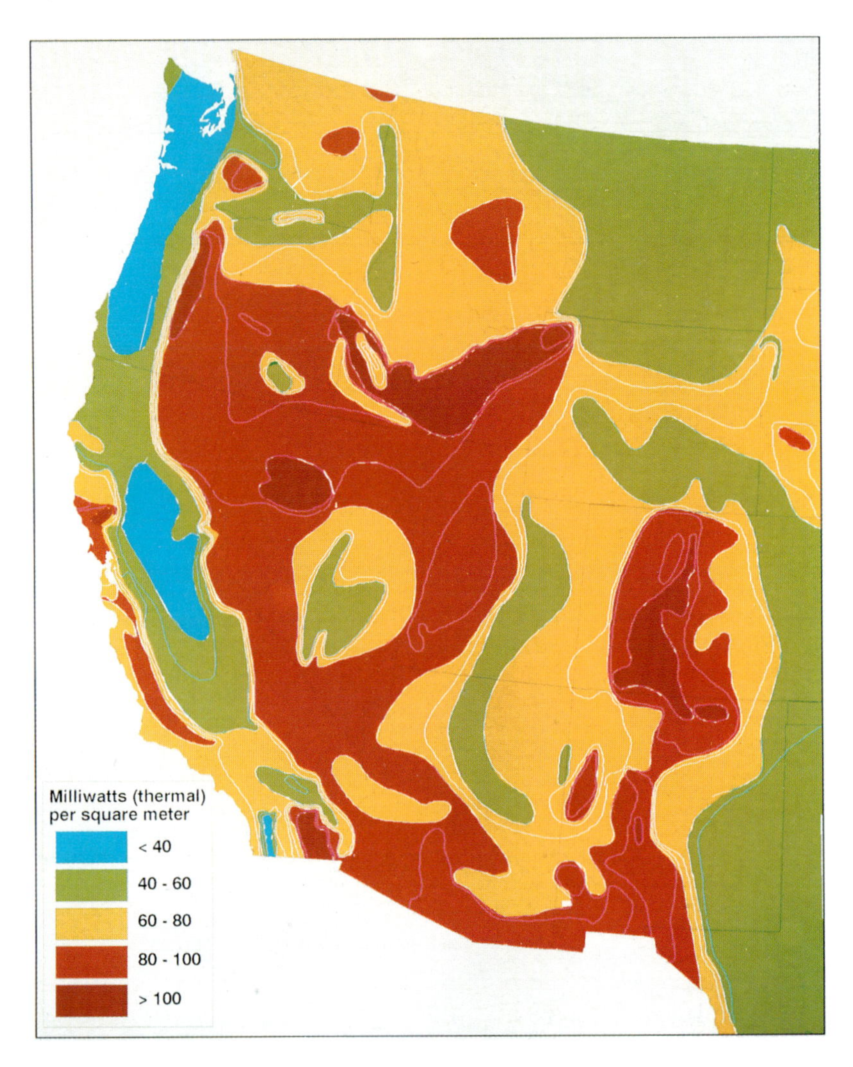

Notes: Heat flow contours are patterned in intervals of 20 milliwatts per square meter (mW/m^2). Highest temperatures will be associated with areas that have both high heat flow and rock strata with high thermal conductivity. The area along the San Andreas fault has moderate to high heat flow. In fact, The Geysers geothermal system is associated with the tectonic effects of the San Andreas fault system. The Sierra Nevada Mountains are notable as one of the lowest heat flow and crustal temperature areas on earth. A *milliwatt (thermal)* is a unit of power in the metric system, expressed in terms of energy per second. See "Watt (Thermal)" in the Glossary.

Source: Modified after the *Geothermal Map of North America*, prepared as part of the Geological Society of North America Decade of North America Geology (DNAG), from Blackwell, D.D., and Steel, J.L., *Mean Temperature in the Crust of the United States for Hot Dry Rock Resource Evaluation* (Southern Methodist University, May 1990), pp. 6-8.

The quantities of each resource category described below are not static. The discovery of additional geothermal fields is expected through the implementation of improved exploration techniques and increased expenditures. As recovery technologies and competitive advantages improve, previously uneconomic resources can become reserves.

Hydrothermal Resources

The most comprehensive assessments of domestic hydrothermal resources were published by the U.S. Geological Survey (USGS) in Circulars 790 and 892.[82,83] The USGS estimated a total accessible resource base consisting of identified and undiscovered

[82]Muffler, L.J.P., editor, *Assessment of Geothermal Resources of the United States—1978*, U.S. Geological Survey Circular 790, 1979, p. 18.

[83]Reed, M.J., editor, *Assessment of Low-temperature Geothermal Resources of the United States—1982*, U.S. Geological Survey Circular 892, 1982, p. 1.

Figure A8. Stages in the Discovery and Development of Geothermal Resources

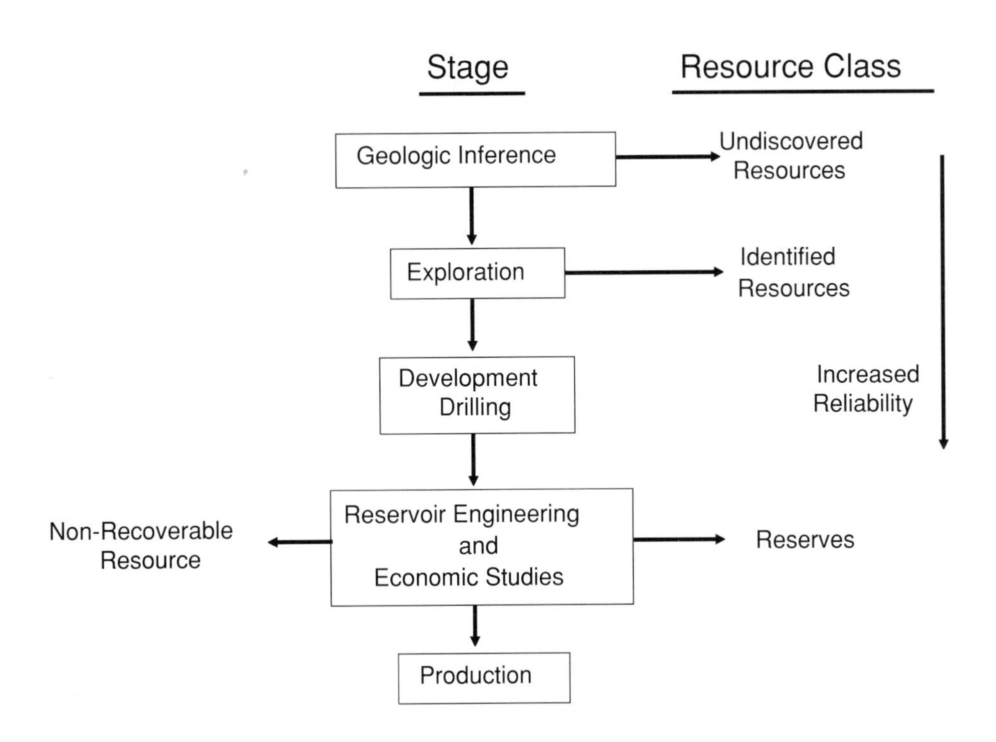

Definitions of Resource Classes[a]

Reserves: that portion of identified resources that can be produced legally at a cost competitive with other commercial energy sources at present. Identified Resources: those resources discovered through drilling, or by geologic, geochemical, or geophysical observation. Undiscovered Resources: resources that have been largely estimated on the basis of broad geologic inference.

[a]Definitions modified from Muffler, C.J.P, editor, *Assessment of Geothermal Resources of the United States—1978*, U.S. Geological Survey Circular 790 (Washington, DC, 1979), p. 4.

components down to a depth of 3 kilometers. Identified resources were discovered through drilling, or by geologic, geothermal, or geophysical observation. Undiscovered resource estimates are based on broad geologic inference. The estimate of total accessible resources, however, did not include those identified systems occurring under lands withdrawn from commercial development, such as Lassen Volcanic National Park and Yellowstone National Park.

Estimates of energy reported by the USGS in Circular 790 for hydrothermal reservoirs with temperatures equal or greater than 90°C are presented in Table A3. Thermal energy contained in reservoirs is estimated at 1,650 quadrillion Btu for identified resources and between 5,900 and 10,000 quadrillion Btu for undiscovered resources. Thermal energy recoverable at the wellhead is estimated at 400 quadrillion Btu for identified resources and 2,000 for undiscovered resources. The USGS estimated electrical energy potentially obtainable from only those hydrothermal systems equal to or greater than 150°C in temperature. Identified resources were estimated at 23,000 (plus or minus 3,400) megawatts capacity for each year of a production lifetime of 30 years. An additional 72,000 to 127,000 megawatts for 30 years was estimated as potentially available from undiscovered resources. Estimates of electrical energy were made without consideration of the economic and market factors that would influence the capability of producers to generate electricity from these resources.

Energy is lost at each step as thermal energy is recovered from the reservoir and then turned into mechanical energy to generate electricity. A 25 percent "geothermal recovery rate" and a capacity factor of approximately 85 percent were applied by the USGS in the estimation of electrical energy obtainable from hydrothermal reservoirs. The recovery rates and capacity factors could actually vary for different reservoirs due to the reservoir characteristics and the reinjection of fluids into the reservoir.

The Bonneville Power Administration (BPA) sponsored a resource assessment for hydrothermal resources in the Cascade region of northwestern United States that included fields which had not been previously reported. Resources were estimated by using a methodology similar to that employed by the USGS. Unlike the USGS study, however, the BPA assessment included areas such as national parks and wilderness areas. Electrical energy potentially obtainable from identified and undiscovered hydrothermal systems in the Cascades with temperatures equal to or greater than 150°C was estimated by the BPA to be at least 185,000 megawatts for 30 years (Table A3). The study assumed that the entire Cascade Range between California and the border with Canada is underlain by a resource between 40 and 60 kilometers in width.[84] This challengeable assumption gives rise to resource estimates that are considerably higher than other published assessments.

Hydrothermal Reserves

Reserves are considered to be that portion of identified hydrothermal resources from which electricity can be produced at a cost competitive with other commercial energy sources at present. Reserves have not been distinguished in the resource assessments described above. The University of Utah Research Institute, in a recent unpublished study completed for the DOE's Office of Conservation and Renewable Energy, estimated hydrothermal reserves at 5,000 megawatts of electric capacity, sustainable for at least 30 years at a busbar cost of 5.5 cents per kilowatthour.[85] Detailed drill hole assessments for all fields were not available for this study; therefore, estimates for some fields were developed on the basis of expert judgment. These findings could be examined in conjunction with information presented in Chapter 3 and in Appendix B. This resource estimate is compared to estimates of potential generation in Chapter 3.

Geopressured, Hot Dry Rock, and Magma Resources

Obtainable energy in the form of electricity has not been estimated for geopressured, hot dry rock, and magma systems because the feasibility to commercially produce electricity from these sources has not been established. The USGS estimated the thermal energy potentially recoverable from geopressured fields, based on uncertain recovery factors, at 430 to 4,400 quadrillion Btu. Recovery techniques are less certain for hot dry rock and magma systems; these resources are usually estimated as the amount of energy contained within an accessible depth of the crust. The energy contained within identified and undiscovered hot dry rock systems accessible to 3 kilometers is estimated at 3.3 million quadrillion Btu.[86] Heat flows greater than 100 mW/m^2 indicate areas of greatest potential (Figure A7). Total resources for magma systems accessible to 10 kilometers are estimated at 100 million quadrillion Btu.[87]

[84]Bloomquist, H.G., Black, G.L., Parker, D.S., Sifford, A., Simpson, S.J., and Street, L.V., *Evaluation and Ranking of Geothermal Resources for Electrical Offset in Idaho, Montana, Oregon, and Washington*, Bonneville Power Administration Report, DOE/BP13609, Volume I, Table 3.2, p. 75.

[85]Wright, P.M., University of Utah Research Institute, oral communication to W. Szymanski, EIA, April 15, 1991.

[86]Tester, J.W., and Herzog, H.J., *Economic Predictions for Heat Mining: A Review and Analysis of Hot Dry Rock (HDR) Geothermal Energy Technology*, Energy Laboratory, Massachusetts Institute of Technology, MIT-EL 90-001, 1990, p. 1. Report prepared for the Geothermal Technology Division, U.S. Department of Energy.

[87]Muffler, L.J.P., editor, *Assessment of Geothermal Resources of the United States—1978*, U.S. Geological Survey Circular 790, 1979, Table 20, p. 157.

Table A3. Estimates of Hydrothermal Resources

	Reservoir Thermal Energy (Quads)	Electrical Energy Installed Capacity (Megawatts over 30 Years)
U.S. Geological Survey Assessment of Western United States, Including Hawaii and Alaska[a]		
Identified Resources[b]		
Vapor-dominated	100	1,610[c]
Hot-water >150°C	850	21,000[d]
Hot-water 90°-150°C[b]	700	NE[e]
Total	**1,650**	**23,000[f]**
Undiscovered Resources		
Vapor-dominated and hot-water >150°C	2,800-4,900	72,000-127,000
Hot-water 90°-150°C	3,100-5,200	NE[e]
Total	**5,900-10,100**	**95,000-150,000**
Bonneville Power Administration Assessment of the Cascades Region of Oregon and Washington[g]		
Identified and Undiscovered Resources		
Vapor-dominated and hot-water >150°C	11,750-17,620[h]	185,000-280,000[h]

[a]Muffler, L.J.P., editor, *Assessment of Geothermal Resources of the United States—1978*, U.S. Geological Survey Circular 790 (1979), Table 4, p. 41.

[b]Estimates for identified resource are mean values.

[c]Excludes an estimated 1,240 MW in Yellowstone National Park.

[d]Excludes reservoirs in Cascades and 47 MW in national parks.

[e]Not estimated.

[f]Totals may not equal sum of components due to independent rounding and assessment methodologies.

[g]Bloomquist, R.G., Black, G.L., Parker, D.S., Sifford, A., Simpson, S.J., and Street, L.V., *Evaluation and Ranking of Geothermal Resources for Electrical Offset in Idaho, Montana, Oregon and Washington,* Bonneville Power Administration Report, DOE/BP13609, Volume I, Table 3.2, p. 75.

[h]Includes national parks and wilderness areas.

DOE Cost-Shared Drilling Resumes in Cascades

The fifth temperature gradient hole has been drilled in the Cascade Range of Oregon under the DOE/industry cost-shared Cascades Deep Thermal Gradient Drilling Program. Initiated in 1985, this program is designed to characterize the deep hydrothermal resource of the Cascades volcanic region and to develop analytical and interpretive tools for industry use in locating and evaluating geothermal reservoirs within young volcanic regions.

The fifth hole, completed in the summer of 1990, was drilled near the Santiam Pass in the Deschutes National Forest by Oxbow Geothermal Co. Previous holes were located on the northern and southern flanks of the Newberry Caldera, on the north slope of Mt. Jefferson near Breitenbush Hot Springs, and on the southeastern slopes of the Crater Lake caldera.

The volcanic region has long been suspected of containing considerable geothermal potential, as evidenced by recent volcanism—e.g., Mount St. Helens—and other thermal expressions. However, there are few known surface manifestations of geothermal energy in spite of the obvious occurrence of heat sources. One possible explanation is that the downward percolation of the extensive regional cold groundwater system forms a so-called "rain curtain" that suppresses surface evidence of the underlying hydrothermal systems.

While a number of holes have been drilled in the area, few have been of sufficient depth to adequately evaluate the temperature and hydrological conditions beneath the cold water zone. In order to support expansion of geothermal development into new areas, the cost-shared program was initiated in this potentially fruitful area to obtain core samples in specifically chosen areas and downhole well logs.

The data obtained have been studied extensively by both the companies involved and DOE-funded researchers. The University of Utah Research Institute is responsible for studies on DOE's share of the core samples and project data. The studies have included lithologic logging of the core, hydrothermal alteration studies, analysis of the geophysical well logs, and physical and chemical measurements on the core samples. The results derived have been placed in the public domain through papers presented in the annual Geothermal Resources Council Transactions. Core samples also are available to the public at UURI in Salt Lake City.

The new 3,046-foot hole at the Santiam Pass, the westernmost of the holes, is on the axis of the Cascades where the highest temperatures are expected. The hole will remain open for research through September 1991. Interested researchers should contact Brittian Hill, Geoscience Department, Oregon State University, Corvallis, Oregon, 97331-5506, or by phone at (503) 737-1201, to coordinate studies.

Source: Excerpted from U.S. Department of Energy, *Geothermal Progress Monitor*, No. 12 (December 1990), pp. 42-43.

Appendix B

Sandia National Laboratory Study: Supply of Geothermal Power from Hydrothermal Sources

Introduction

This appendix provides a description of the methodology for calculating current and projected geothermal electric power estimates used as part of a hydrothermal sources study that was commissioned by the Energy Information Administration and the DOE Office of Conservation and Renewable Energy.[88] This appendix also summarizes the study's electric power cost estimates over the 40-year forecast horizon obtained from four scenarios developed in the study: (1) current technology, current identified resources (base case); (2) current technology, identified and estimated unidentified resources; (3) improved technology, current identified resources; and (4) improved technology, identified and estimated unidentified resources. Results for the base case and the most optimistic case (improved technology, identified and estimated unidentified resources) are presented as projected generation capacities or so-called "supply curves" in Table B1.

Resource assessments of geothermal power were based on the following sources:

- U.S. Geological Survey (USGS) estimates of electric capacity potentially obtainable from hydrothermal reservoirs

- National Oceanic and Atmospheric Administration (NOAA) maps of geothermal resources for the States of Arizona, California, Nevada, New Mexico, Washington, Oregon, Colorado, Montana, Idaho, and Utah

- Bonneville Power Administration's resource assessment of geothermal electric power in the Pacific Northwest[89]

- Other published reports and scientific papers

- Personal knowledge on the part of the study team and their contacts in the geothermal energy business.

Using USGS Circular 790 as a starting point, lists were made of all the known or identified resources in the States of California, Nevada, Oregon, Washington, Idaho, and Utah for which the USGS had indicated any potential for generating power. The USGS considered a resource identified only if it had some surface manifestation such as hot springs, fumaroles, or geysers, or if a well had been drilled into the resource. In addition to the USGS criteria, the study defined areas exhibiting active volcanism or other thermal features, such as above average thermal gradients. Estimation of identified resources included a determination of the amount of power currently on line at each resource or the amount of power which would be on line within the next 5 years (end of 1995). For power to be assumed to be on line in 5 years, the study required that a power plant be under construction or a firm power sales agreement with permits for plant construction be in effect. The analysis estimated how much hydrothermal energy could be physically available for sale, not how much would be sold. The analysis also looked at the status of resource exploration. If active exploration was underway, the developer was contacted and asked how much power could be on line in 20 and 40 years, assuming power sales agreements were possible. A resource was assumed to be available for development if land was leased and an active program of exploration was underway.

The USGS study limited resources capable of electric power generation to those above 150°C. The Sandia study developed resource assessment data on low-

[88]Petty, S., Livesay, B.J., and Geyer, J., *Supply of Geothermal Power from Hydrothermal Sources: A Study of the Cost of Power Over Time*, prepared for the U.S. Department of Energy (Sandia National Laboratory, 1991), pp. 8-12 (Draft).

[89]Bloomquist, R.G., Black, G.L., Parker, D.S., Sifford, A., Simpson, S.J., and Street, L.V., *Evaluation and Ranking of Geothermal Resources for Electrical Generation or Electrical Offset in Idaho, Montana, Oregon, and Washington*, Vol. I and II (Bonneville Power Administration, 1985).

Table B1. Projected Capacities of Hydrothermal Resource Sites in the Western United States and Hawaii
(Megawatts Electric)

State/Resource Site	Capacity				
	Base Case			Improved Technology and Accelerated Exploration Case	
	1995	2010	2030	2010	2030
Arizona					
Power Ranches	-	190	475	380	950
Total	**0**	**190**	**475**	**380**	**950**
California					
Brawley	-	150	300	350	640
Buckeye HS	-	250	635	500	1,270
Clear Lake	-	500	900	500	900
Coso/China Lake	250	650	650	650	1,000
East Mesa	107	360	360	360	500
Geysers	2,000	2,000	2,000	2,000	2,000
Glamis	-	275	680	275	680
Heber Geothermal	100	250	250	250	500
Kelly Hot Springs	-	300	760	1,180	3,000
Lassen	-	116	250	100	350
Long Valley (LT)	20	250	500	350	750
Long Valley (HT)	-	500	1,600	500	1,600
Medicine Lake	25	500	2,000	750	3,000
Niland (see Salton Sea)					
Randsburg	-	25	85	100	250
Routt	-	65	165	130	330
Salton Sea	185	500	1,000	500	3,000
Sespe HS	-	125	330	250	660
Surprise Valley	10	250	500	500	1,490
Wendell	-	250	650	250	650
Westmoreland	-	50	150	150	1,710
Wilbur HS	-	500	1,500	1,000	2,800
Total	**2,697**	**7,866**	**15,265**	**10,645**	**27,080**
Colorado					
Hot Springs Ranch	-	540	1,350	540	1,350
Paradise HS	-	25	100	50	200
Waunita	-	205	515	410	1,030
Total	**0**	**770**	**1,965**	**1,000**	**2,580**
Hawaii					
Kilauea SW Rift	-	50	150	100	300
Puna	3	100	500	200	1,000
Total	**3**	**150**	**650**	**300**	**1,300**
Idaho					
Cove Creek	-	25	100	200	300
Island Park	-	250	1,000	500	2,000
Magic Reservoir	-	360	900	720	1,800
Raft River	5	30	195	250	1,000
Total	**5**	**665**	**2,195**	**1,670**	**5,100**

See footnotes at end of table.

Table B1. Projected Capacities of Hydrothermal Resource Sites in the Western United States and Hawaii (Continued)
(Megawatts Electric)

State/Resource Site	Capacity				
		Base Case		Improved Technology and Accelerated Exploration Case	
	1995	2010	2030	2010	2030
New Mexico					
Rio Grande Rift	-	120	300	240	600
Valles Caldera	-	250	1,000	250	1,000
Total	**0**	**370**	**1,300**	**490**	**1,600**
Nevada[a]					
Beowawe	16	50	130	150	250
Desert Peak	20	100	500	250	1,000
Dixie Valley	60	250	500	250	500
Total	**96**	**400**	**1,130**	**650**	**1,750**
Oregon					
3 Creeks Butte	-	100	500	500	2,000
Alvord Desert	-	100	200	100	575
Klamath Falls	-	100	500	500	2,000
Newberry	-	100	1,000	250	1,500
Vale	2	425	1,062	850	2,000
Total	**2**	**825**	**3,262**	**2,200**	**8,075**
Utah					
Cove Fort	10	150	500	300	1,000
Roosevelt Springs	30	250	500	250	500
Total	**40**	**400**	**1,000**	**550**	**1,500**
Washington					
Mt. Baker	-	25	200	50	400
Total	**0**	**25**	**200**	**50**	**400**
Grand Total	**2,843**	**11,661**	**27,442**	**17,935**	**50,335**

[a]Stillwater/Soda Lake and Steamboat Springs were assumed to be similar to other sites, and were included under the other sites.
Source: Petty, S., Livesay, B.J., and Geyer, J., *Supply of Geothermal Power from Hydrothermal Sources: A Study of the Cost of Power Over Time*, prepared for the U.S. Department of Energy (Sandia National Laboratory, 1991), Tables 1 and 2.

to moderate-temperature resources to estimate the potential for power production from those lesser quality resources. This incremental low-temperature resource potential is defined as the unidentified resource base for the purposes of the Sandia study. Low-temperature resource estimates were expanded from data given in Circular 790 by converting quantities to an electric equivalent.

The Bonneville Power Administration (BPA) study of the Pacific Northwest was utilized to augment data on the resources in the Cascades. Data collected on recent drilling activity, volcanism, and the existence of high heat flow anomalies were used by BPA to identify many more resources than were cited by the USGS. Unidentified resources were estimated as the lesser of either the BPA estimates or twice the current exploration estimate. Estimated unidentified resources were assumed to be the greater of USGS estimates, developer estimates, or the consultant's judgement. Other unidentified resources were assumed to be equal to the USGS estimate, or 50 percent greater than the current exploration estimate.

Resources estimated in the preceding manner were subsequently checked against State geothermal maps prepared under the auspices of the National Oceanic and Atmospheric Administration (NOAA). Several additional resources were added to the lists from these maps. Information on temperature, depth, well flow, geology, and fluid chemistry was compiled for each resource, using published reports, State geothermal maps, and contacts with resource developers. Estimates of the recoverable power were also checked against these sources.

Electric Power Cost Derivation

The Sandia study's electric power cost estimates, which are calculated on a revenue requirements basis and reported as levelized busbar costs in mid-1987 dollars, reflect only those plant and operating cost elements affected by geothermal resource development considerations. Such considerations include reservoir geology, key physical resource characteristics, and the pace and extent of technology improvements. The costs associated with transmitting electricity to the utility grid, as well as other institutional and power marketing factors, were not factored into the cost estimates.

The cost of producing electric power from each resource of the electric power resource base was calculated using a microcomputer model, "Impacts of Research and Development on Cost and Geothermal Power" (IMGEO), Version 3.05. This model was originally developed by the Sandia research team to help analyze how technology improvement from research and development would affect the cost of geothermal power. The IMGEO model has three basic data input categories: baseline technology inputs, site-case inputs, and financial assumption inputs. The IMGEO model estimates the range of costs associated with development of geothermal resources by using a most likely estimated value for critical reservoir parameters such as depth, temperature, and well flow rate, and a worst case estimate for each of these parameters.

The baseline technology inputs consist of cost and engineering performance factors (as of early 1986) associated with liquid-dominated hydrothermal electric subsystems. Cost and performance factors for exploration and confirmation activity, well, field piping, flow testing, and power plant components were developed using actual cost data and cost estimates based on industry experience, available theory, and conceptual designs. The IMGEO model also includes technology improvement "levers" to model percentage changes in technology performance and costs which result from attaining one or more of the objectives.

The site-case inputs consist of reservoir characteristics and plant type designations that are grouped into eight representative site-cases. To derive the eight site-cases, geothermal resources were divided into four regions roughly equivalent to the USGS geothermal regions: (1) the Imperial Valley in California, (2) the Basin and Range region, (3) the Cascades in Oregon and California, and (4) young volcanics. The young volcanics regions includes all hydrothermal resources associated with recent volcanism other than the Cascades. Cost factors such as fluid chemistry, number of dry holes per producer, number of injectors per producer, rate of well workover, and cost of well workover were tied to these regional designations. For the Sandia study, each resource was assigned to one of these regions. Some resource areas, particularly those in Colorado, Montana, and Idaho, did not fit into these regions and were categorized as part of the region with closest similar geology.

For each of the four regions, a high-temperature (>200°C) flash steam case and a low-temperature (<200°C) binary case were defined, thus resulting in eight site-cases. Each resource was then assigned to either the flash steam or binary development default input for the appropriate region. Three factors—temperature, depth, and flow rate per well—were extracted from the resource data base for each resource as input to IMGEO. The IMGEO model then calculated system performance and cost estimates using the technology and site-case reservoir data.

To derive the levelized busbar cost of electricity for each resource, a set of financial input assumptions were applied. The financial assumptions used by IMGEO reflect a utility financing case recommended in the 1987 Electric Power Research Institute (EPRI) Technical Assessment Guide for analyzing renewable energy project costs. The utility financing case was used by the DOE Office of Conservation and Renewable Energy for cross-technology cost analyses. The set of financial factors used in the IMGEO model are summarized in Table B2.

The cost of transmitting geothermal electric power over long distances to the electric power grid was not included in the calculations of power cost for the Sandia study. The changes in the utility transmission grid in the future are difficult to project. The cost of transmission lines to tie individual resources into the electric power grid needs to be estimated and included in calculating the cost of power production. An internal analysis by DOE, citing an average transmission distance of 700 miles from geothermal site to end-use

Table B2. Financial Assumptions of the Sandia Study

Factor	Value
Cost Reporting Year, Average	1987.5
Cost Reporting Year, Wells	1990.5
Cost Reporting Year, Power Plant	1986.0
Years to Construct Power Plant	2.5
Levelized Annual Capacity Factor	0.80
Cost Basis: Overnight Construction, AFUDC Not Included in Model Costs	
Allowance for Interest During Construction	1.081
General and Fuel Inflation Rate	0.06
Discount Rate = Weighted Cost of Capital (For Levelization in Current Dollars) (For Levelization in Constant Dollars, Use: 1 - (1.1249)/(1.06))	0.1249
Levelized Annual Capital Charge Rate for Calculations in Current Dollars (Includes Amortization, Income Taxes, Taxes Incentives, Property Tax, and General Property Insurance)	0.1683
Current Dollars Cost/Constant Dollars Cost	1.747961
General and Fuel Cost Levelization Factor	1.748
Book Life of Project, Years	30
Tax Life, Years	15
Combined Federal and State Income Tax Rate	0.38
Investment Tax Credit Rate	0.00
Property Tax and Insurance (Accounting Method: Normalization) (Accelerated Depreciation: Double Declining Balance)	0.02
Geothermal Production Field Special Financial Factors	
Royalty Rate	0.10
Severance Tax	0.04
Percent Depletion Allowance	0.15
Intangible Fraction of Well Cost	0.75

Source: Petty, S., Livesay, B.J., and Geyer, J., *Supply of Geothermal Power from Hydrothermal Sources: A Study of the Cost of Power Over Time,* prepared for the U.S. Department of Energy (Sandia National Laboratory, 1991), pp. 8-12 (Draft).

area, estimated the average capacity cost at 1.1 cents per kilowatthour (1985 dollars). For a distance of 100 miles, the cost was estimated at 0.2 cents per kilowatthour. These costs do not account for cost escalation, actual line length, terrain, population density, or type of transmission structure.

Estimates of Current Costs of Electric Power

The supply of geothermal power from presently identified resources ranges in cost from about 2 cents per kilowatthour up to 25 cents per kilowatthour (Figure B1), assuming the current rate of technology improvement continues. This is the business as usual scenario, with few developers taking the risk of trying new technologies and a limited budget for government-sponsored research. Figure B2 shows a similar plot with unidentified resources included in the total.

An upper cost limit of 2.5 cents per kilowatthour was specified. Some upper limit had to be chosen to reduce the number of potential resources for consideration due to the lack of data on these less economic resources and the amount of time necessary to process these scarce data. However, it should be understood that this is an artificial limit. Resources which are currently costly to develop are generally much more amenable to improvement in cost through research efforts.

Figure B1. Potential Supply of Geothermal Power: Current Technology, Identified Resources

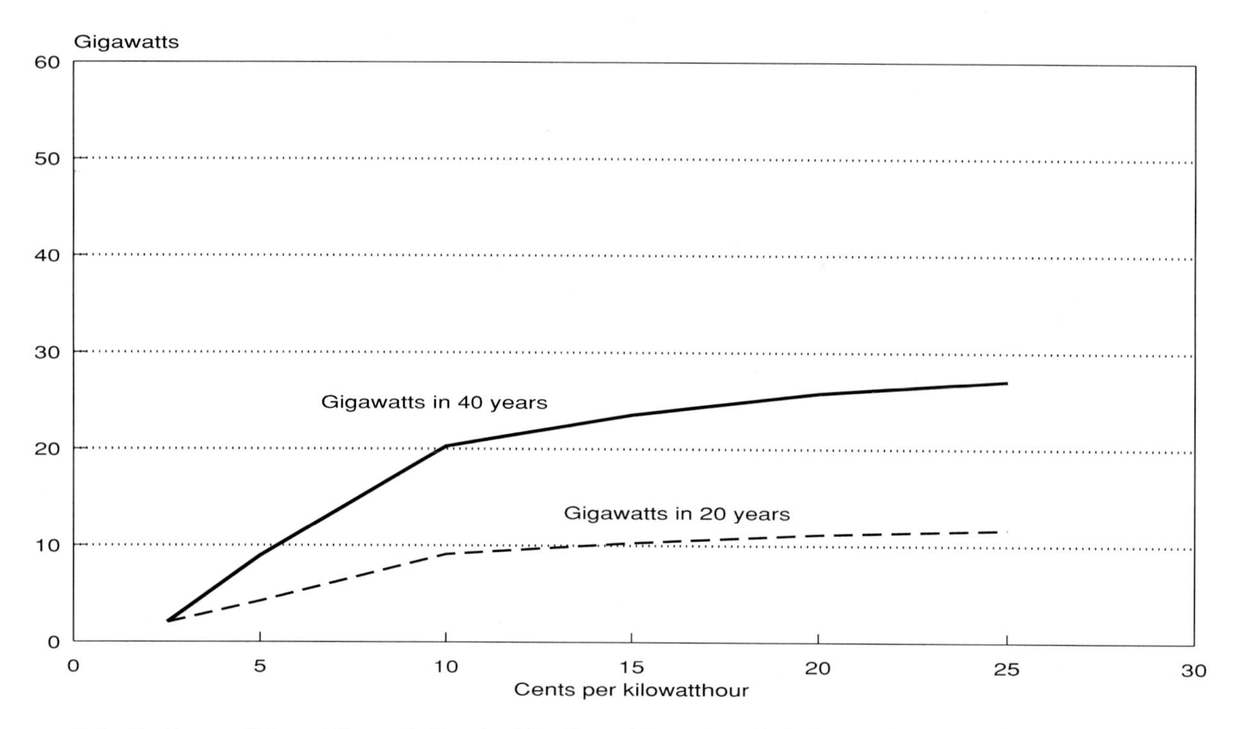

Source: Petty, S., Livesay, B.J., and Geyer, J., *Supply of Geothermal Power from Hydrothermal Sources: A Study of the Cost of Power Over Time,* prepared for the U.S. Department of Energy (Sandia National Laboratory, 1991), pp. 8-12 (Draft).

Figure B2. Potential Supply of Geothermal Power: Current Technology, Identified and Estimated Unidentified Resources

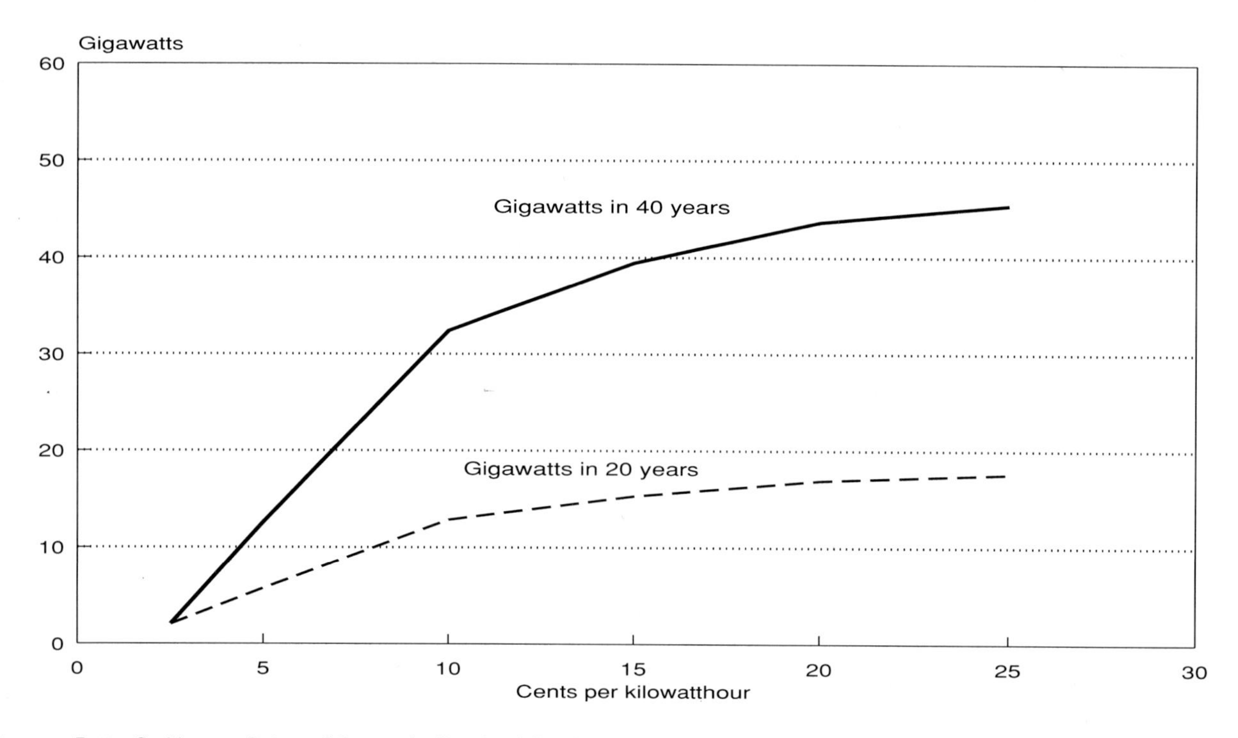

Source: Petty, S., Livesay, B.J., and Geyer, J., *Supply of Geothermal Power from Hydrothermal Sources: A Study of the Cost of Power Over Time,* prepared for the U.S. Department of Energy (Sandia National Laboratory, 1991), pp. 8-12 (Draft).

The amount of available electricity capacity levels off with increasing cost of power. This characteristic is related to the scarcity of data on less explored resources and is, therefore, in part an artifact of data availability. The USGS and other published reports concentrate on identified resources. Data are most likely to be available on resources which are more economic to produce. Deep resources, resources with low temperatures, and resources with low potential productivity are not likely to be explored by either researchers or developers until most likely prospects, such as the Cascades, are studied. When the unidentified resources were included in the capacity-cost plot, this flattening was reduced. Further study of unidentified resources could provide insight into the actual relationship between the electric power resource base and the costs to produce that power.

Estimates of Costs of Electric Power with Technology Improvements

An additional set of estimates were developed to simulate the cost of power reflecting assumed technology improvements over the next 40 years. IMGEO was used for these calculations, assuming the research and development achievements listed in Table B3.

These research goals represent the authors' estimation of the maximum improvement possible using existing methods for each of the chosen factors. No major breakthroughs in drilling, testing, or exploration were considered. The improvements are realized through evolutionary change in currently available technology. The assumption was made that the maximum improvement could be reached by 2030 if this was the goal of the geothermal industry and government. Straight-line interpolation was used between present cost of power and the cost changes from the research and development impacts.

Table B3. Technology Improvement Assumptions

Factor Addressed	Percentage Improvement Assumed
Wildcat Success Ratio	20 percent greater
Confirmation Success Ratio	25 percent greater
Testing Costs, Confirmation	25 percent less
Dry Holes/Producer	15 percent fewer
Testing Costs/Producer	25 percent less
Base Cost, Ave. Well	20 percent less
Capital Cost, Deep Well Pump	25 percent less
O&M Cost, Deep Well Pump	20 percent less
Workover Interval, Production	50 percent shorter
Workover Interval, Injection	50 percent shorter
Flash Plant Efficiency	5 percent better
Flash Plant, Capital Cost	5 percent less
Binary Plant Efficiency	20 percent greater
Binary Plant Capital Cost	24 percent less
Removal of Solids, Capital Cost	10 percent less
Removal of Solids, O&M Cost	20 percent less
H_2S Treatment, Capital Cost	20 percent less

Source: Petty, S., Livesay, B.J., and Geyer, J., *Supply of Geothermal Power from Hydrothermal Sources: A Study of the Cost of Power Over Time,* prepared for the U.S. Department of Energy (Sandia National Laboratory, 1991), p. 11 (Draft).

Figure B3 shows the supply of power at various costs with technology improvement for the identified resources in the western United States and Hawaii. Figure B4 shows a similar plot with unidentified resources included. The largest percentage impact in cost improvement was found for the resources with the highest cost to produce. This is as expected and suggests that further study of these high-cost resources could increase the resource base.

Figure B3. Potential Supply of Geothermal Power: Technology Improvement Assumptions, Identified Resources

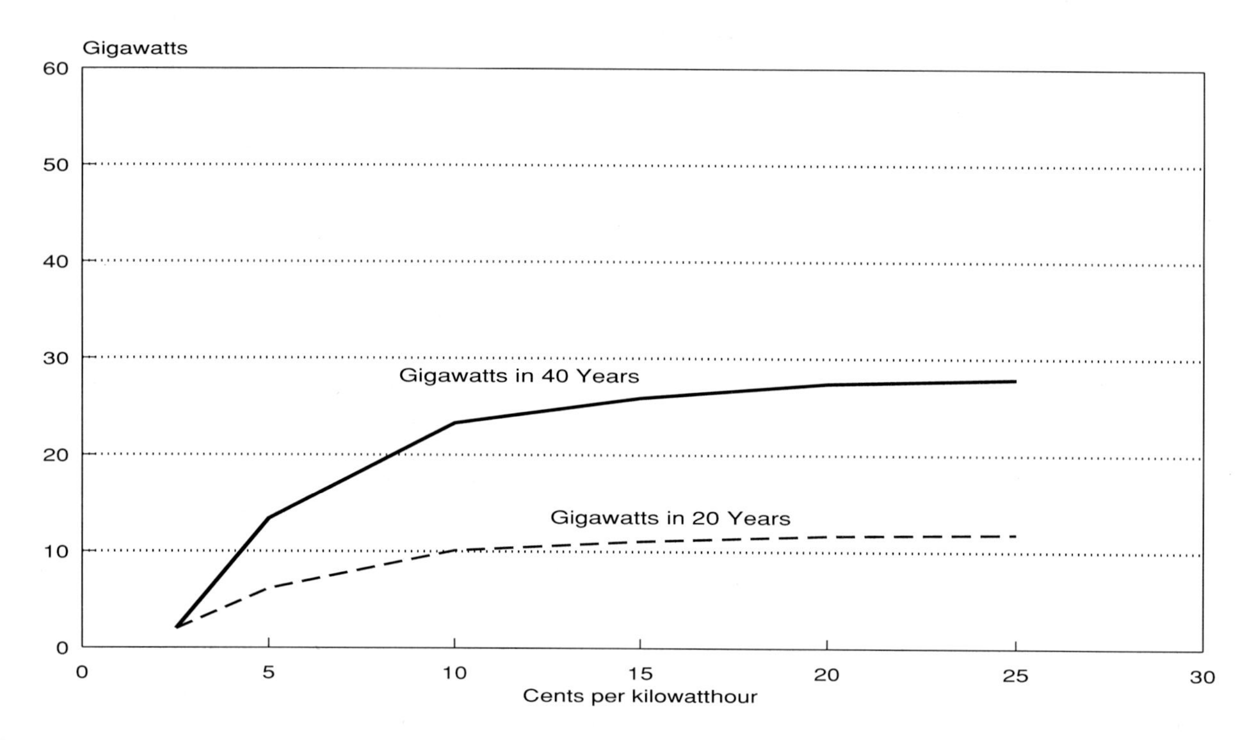

Source: Petty, S., Livesay, B.J., and Geyer, J., *Supply of Geothermal Power from Hydrothermal Sources: A Study of the Cost of Power Over Time,* prepared for the U.S. Department of Energy (Sandia National Laboratory, 1991), pp. 8-12 (Draft).

Figure B4. Potential Supply of Geothermal Power: Technology Improvement Assumptions, Identified and Estimated Unidentified Resources

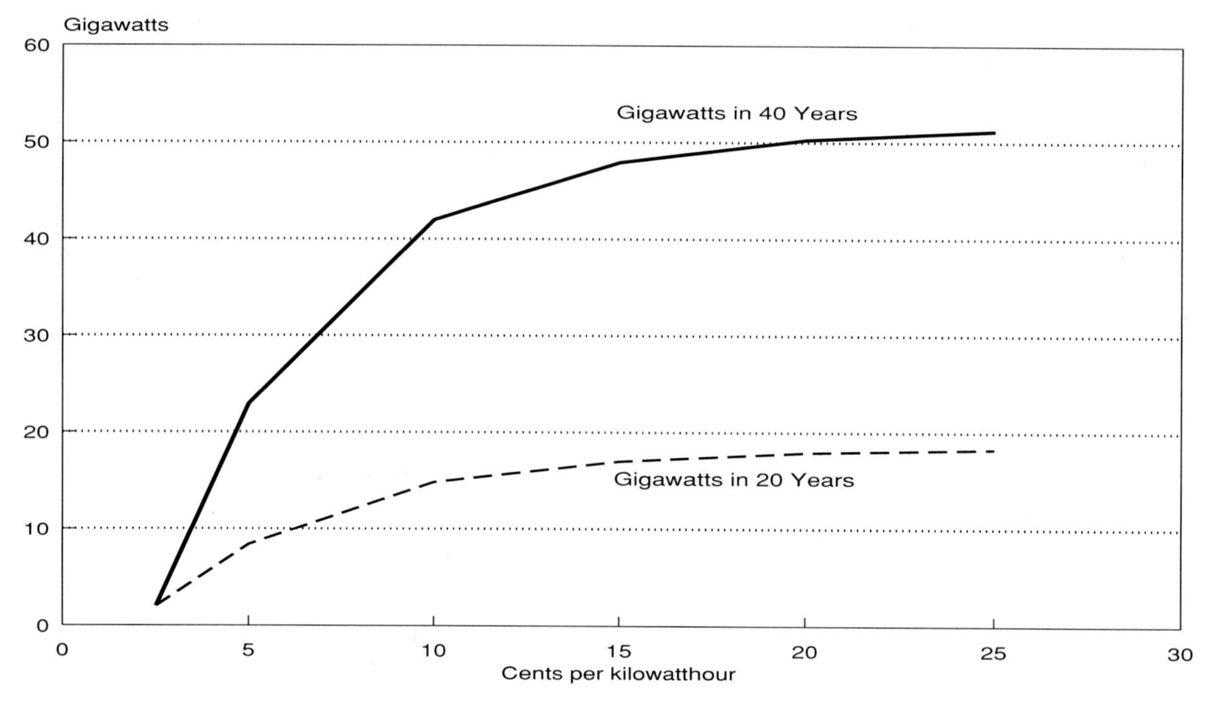

Source: Petty, S., Livesay, B.J., and Geyer, J., *Supply of Geothermal Power from Hydrothermal Sources: A Study of the Cost of Power Over Time,* prepared for the U.S. Department of Energy (Sandia National Laboratory, 1991), pp. 8-12 (Draft).

Appendix C

Descriptions of Liquid-Dominated Geothermal Power Plants

Coso Hot Springs, Inyo County, California

- **Navy 1**—84 megawatts net installed capacity for three units (net), owned by Caithness Corporation, operated by California Energy, resource temperature 400-650°F double flash, installed cost $3,400 per kilowatt, on line since 1987. Electricity sold to Southern California Edison.

- **Navy 2**—84 megawatts net, owned by Caithness Group, California Energy, and Dominion Energy, operated by California Energy, resource temperature 400-650°F, double flash. Electricity sold to Southern California Edison.

- **BLM East**—56 megawatts net, for two units owned by Caithness Corporation, California Energy, and Dominion Energy, resource temperature, 400-650°F, operated by California Energy, double-flash system. Electricity sold to Southern California Edison.

- **BLM West**—28 megawatts net plant owned by California Energy Company. Electricity sold to Southern California Edison.

East Mesa, Imperial County, California

- **McCabe 1**—13 megawatts net, owned by GEO (50 percent) and Mission Energy (50 percent), operated by Magma Power, binary, resource temperature 350°F. Electricity sold to Southern California Edison.

- **Ormesa 1**—24 megawatts net, owned by Ormat (50 percent) and LCF Financial (50 percent), resource temperature 320°F, binary, uses Rankine cycle technology, world's largest modular binary geothermal power plant. Electricity sold to Southern California Edison.

- **Ormesa 1E**—4 megawatts net. Electricity sold to Southern California Edison.

- **Ormesa 1H**—6 megawatts net. Electricity sold to Southern California Edison.

- **Ormesa 2**—17 megawatts net, owned by Constellation Development (50 percent) and Tricon (50 percent), operated by Ormat, resource temperature 355°F, binary, capital cost $70 million. Electricity sold to Southern California Edison.

- **Gem 2**—18 megawatts gross, owned by GEO (50 percent) and Mission Energy (50 percent), operated by GEO, double flash. Electricity sold to Southern California Edison.

- **GEM 3**—19 megawatts net.

Heber, Imperial County, California

- **Heber Double Flash**—47 megawatts net, owned by Centennial and ERC Corporations, resource temperature 360°F, double-flash system, built in 1985. Electricity sold to Southern California Edison. The estimated capacity of the Heber reservoir was reduced from 117 megawatts in 1989 to 47 megawatts in 1990.

Mammoth Lake, Mono County, California

- **Mammoth Pacific 1**—10 megawatts net, owned by Pacific Energy (Partnership of Dravo Corporation and Energy Centennial Energy), installed cost $1,250 per kilowatt, resource temperature 400°F, binary, on line 99.8 percent of the time. Power sold to Southern California Edison.

Salton Sea, Imperial County, California

- **Vulcan**—32 megawatts net, owned by Magma/ Mission Energy, resource temperature 550°F, double-flash system, cost $72 million, wells directionally drilled. Electricity sold to Southern California Edison.

- **Del Ranch**—36 megawatts gross, 34 megawatts net, owned by Magma / Mission Energy, double-flash. Electricity sold to Southern California Edison.

- **Elmore 1**—36 megawatts net, double-flash system. Electricity sold to Southern California Edison.
- **Leathers 1**—36 megawatts net, owned by Magma Power and Mission Energy, double-flash system, plant cost $100 million, sells power to Southern California Edison under Standard Offer Number Four contract. Electricity sold to Southern California Edison.
- **Salton Sea 1**—10 megawatts net, single flash system, owned by Earth Energy (UNOCAL subsidiary).
- **Salton Sea 2**—18 megawatts net, single flash system, owned by Earth Energy (UNOCAL subsidiary).
- **Salton Sea 3**—51 megawatts net, owned by Desert Power (subsidiary of UNOCAL), double flash, resource temperature 625°F; Desert Power will participate in a joint industry transmission line. Electricity sold to Southern California Edison.

Wendall, Lassen County, California

- **Amadee Geothermal**—1.6 megawatts net, resource temperature 220°F, binary, owned by Trans-Pacific Geothermal Inc. Electricity sold to Pacific Gas & Electric Company.
- **Wineagle Project**—1 megawatt net, binary system owned by Wineagle Developers. Electricity sold to Pacific Gas & Electric Company.
- **Honey Lake**—5 megawatts geothermal, 25 megawatts wood/waste-fired, total plant 30 megawatts, binary system. Electricity sold to Pacific Gas & Electric Company.

Puna, Hawaii

- **Puna**—3.0 megawatts gross, 2.5 megawatts net, single-flash, on line 1981. Decommissioned in 1989 because it had exceeded its 3-year demonstration design life.

Nevada

- **Beowawe, Lander County**—16.6 megawatts net, resource temperature 430°F, double-flash, on line 1985, owned by California Energy and Crescent Valley Geothermal (Southern California Edison subsidiary), operated by Chevron Resources. Electricity sold to Southern California Energy.
- **Desert Peak, Churchill County**—9 megawatts net, double-flash, operating since 1985, resource temperature 435°F, owned by California Energy Company.
- **Dixie Valley, Churchill County**—57 megawatts net, owned by Oxbow Geothermal, double-flash, on line 1988.
- **Empire, Washoe County**—3.1 megawatts net, owned by Ormat and Constellation Development, binary, on line 1987.
- **Soda Lake, Churchill County**—2.7 megawatts net, binary, resource temperature 320°F.
- **Soda Lake II**—13 megawatts net.
- **Steamboat Geothermal No. 1 and No. 1-A**—6.8 megawatts net, owned by Far West Hydroelectric, operated by Ormat, Inc., on line 1986, binary, air-cooled.
- **Steamboat Springs 1**—12.5 megawatts net, owned by Caithness Group/Sequa Corporation, operated by Cathiness, Inc., resource temperature 340°F, single-flash, on line 1988, 84 percent capacity factor, 99 percent availability.
- **Stillwater Geothermal Project, Fallon County**—13 megawatts gross, owned by Constellation Development and Chrysler Capital, resource temperature 310-340°F, binary, air-cooled, 100 percent of brines reinjected.
- **Wabuska, Lyon County**—1.0 megawatts net, resource temperature 225°F, binary, on line 1984, operated and owned by Tad's Enterprises.

Texas

- **Pleasant Bayou, Brazoria County**—1 megawatt, geopressured geothermal demonstration plant. Operated for 12 months; no longer in operation.

Utah

- **Cove Fort Geothermal 1**—2 megawatts gross, flash, combined cycle, on line 1985, owned by City of Provo, developed by Mother Earth Industries.
- **Blundell 1, Roosevelt Springs**—20 megawatts net, single-flash, on line 1984, resource temperature 520°F, availability approximately 90 percent. Electricity sold to Utah Power Division of Pacificorp.

Appendix D

Geothermal Project Developers and Owners

The following list contains those companies that are believed to be currently active in developing geothermal fields in the western United States and Hawaii, or have an ownership or revenue interest in domestic geothermal projects. Major sources of information used to construct this list include: *Power Plays 1989*, Investor Responsibility Research Center; *Independent Energy*; *Power Magazine*; *Geothermal Progress Monitor*, Geothermal Division, U.S. Department of Energy; Membership Roster, Geothermal Resource Council; and direct company contacts.

While this list includes a majority of the companies involved in the development and ownership of geothermal energy resources, entities such as small, newly formed joint ventures or partnerships may not be fully represented.

Anadarko Petroleum Corporation
Geothermal Division, 835 Piner Rd., Suite A
Santa Rosa, CA 95403

Bonneville Pacific Corporation
257 East 2nd South, Suite 800
Salt Lake City, UT 84111

Caithness Corporation
1114 Avenue of the Americas, 35th Floor
New York, NY 10036

California Energy Company
601 California Street, Suite 900
San Francisco, CA 94108

Calpine Corporation
P.O. Box 11279
Santa Rosa, CA 95406

Centennial Energy Corporation
650 California Street, Suite 2250
San Francisco, CA 94108

Coldwater Creek Operator Corporation
1330 N. Dutton Avenue, Suite A
Santa Rosa, CA 95409

Community Energy Alternatives Inc.
Parent Company: Public Service
Enterprise Group
1200 East Ridgewood Ave., 2nd Floor
Ridgewood, NJ 07450

Constellation Energy Inc.
Parent Company: Baltimore Gas &
Electric Company
250 W. Pratt St., 23rd Floor
Baltimore, MD 21201

Chrysler Capital Corporation
Parent Company: Chrysler Corporation
225 High Ridge Road
Stamford, CT 06905

Diablo Executive Group Inc. (The)
P.O. Box 546
Diablo, CA 95428

Dominion Energy Resources
Parent Company: Dominion Resources Inc.
P.O. Box 26532
Richmond, VA 23261

ERC Environmental & Energy Services
Parent Company: ERC International
3211 Jermantown Road
Fairfax, VA 22030

ESI Geothermal
Parent Company: ESI Energy Inc.
1400 Centrepark Blvd., #600
West Palm Beach, FL 33401

Far West Hydroelectric Fund
921 Executive Park Dr., Suite B
Salt Lake City, UT 84117

GeoProducts Corporation
1850 Mt. Diablo Blvd., Suite 300
Walnut Creek, CA 94596

Geothermal Development Associates
251 Ralston Street
Reno, NV 89503

Geothermal Power Company Inc.
1460 West Water Street
Elmira, NY 14905

Geothermal Resources Council
2001 2nd Street, Suite 5
P.O. Box 1350
Davis, CA 95616

Geothermal Resources International Inc.
1825 South Grant St., Suite 900
San Mateo, CA 94402

GeothermEx Inc.
5221 Central Ave., Suite 201
Richmond, CA 94804

Honey Lake Industries Inc.
1800 Apple View Way
Paradise, CA 95969

Harbert Power Group
Parent Company: Harbert
International and Sita, S.
150 Spear Street, No. 1875
San Francisco, CA 94105

LFC Financial Corporation
3 Radnor Corporate Center,
100 Matsonford Rd.
Radnor, PA 10987

Los Angeles Department of Water & Power
P.O. Box 111, Room 1141
Los Angeles, CA 90051

Magma Power Corporation
11770 Bernardo Plaza Court, Suite 366
San Diego, CA 61948

Maxus Energy Corporation
717 N. Harwood Street, 28th Floor
Dallas, TX 75201

Mission Energy Company
Parent Company: Southern California Edison
18872 MacArthur Blvd., Suite 400
Irvine, CA 92715

Mother Earth Industries Inc.
7350 East Evans Road, Suite B
Scottsdale, AZ 85260

Northern California Power Agency
180 Cirby Way
Roseville, CA 95678

Ormat Energy Systems
Parent Company: The Ormat Group
610 East Glendale Ave.
Sparks, NV 89431

Oxbow Geothermal
Parent Company: Oxbow Corporation
333 Elm Street
Dedham, MA 02026

Pacific Energy
Parent Company: Pacific Enterprises
6055 E. Washington #608
Commerce, CA 90040

SAI Geothermal
Parent Company: SAI Engineers Inc.
3030 Patrick Henry Drive
Santa Clara, CA 95054

Santa Fe Geothermal
Parent Company: Santa Fe International Corp.
13455 Noel Road, Suite 1100, Two Galleria
Tower Dallas, TX 75240

Santa Rosa Geothermal Company
Parent Company: Calpine Corp. and
Freeport-McMoran
Partnership
P.O. Box 11279
Santa Rosa, CA 95406

Steam Reserve Corporation
Parent Company: Amax Exploration Inc.
1625 Cole Blvd.
Golden, CO 80401

Tad's Enterprises
P.O. Box 145
Yerington, NV 89447

Thermal Exploration Company
970 E. Main Street, Suite 100
Grass Valley, CA 95945

Trans-Pacific Geothermal Corporation
1330 Broadway, Suite 1525
Oakland, CA 94612

Tricon Leasing Corporation
Parent Company: Bell Atlantic
11720 Beltsville Dr.
Beltsville, MD 20705

True Geothermal
Parent Company: True Oil
P.O. Drawer 2360
Casper, WY 82602

Unocal Corporation
1201 W. Fifth Street
Los Angeles, CA 90017

University of Utah Research Institute
Earth Science Laboratory
391 Chipeta Way, Suite C
Salt Lake City, UT 84108-1295

U.S. Department of Energy
Geothermal Division
1000 Independence Ave., S.W.
Washington, DC 20585

U.S. Energy Corporation
1755 East Plumb Lane, Suite 265A
Reno, NV 89502

Yankee Power Inc.
400 S. Boston #310
Tulsa, OK 74103

Glossary

Aquifer: A subsurface rock unit from which water can be produced.

Arcuate Structure: A geologic formation that is curved or bowed.

Availability Factor: A percentage representing the number of hours a generating unit is available to produce power (regardless of the amount of power) in a given period, compared to the number of hours in the period.

Avoided Costs: The incremental costs of energy and/or capacity, except for the purchase from a qualifying facility, a utility would incur itself in the generation of the energy or its purchase from another source.

Balneology: The body of knowledge dealing with the therapeutic effects of bathing.

Baseload: The minimum amount of electric power delivered or required over a given period of time at a steady rate (See Baseload Plant).

Baseload Plant: A plant, usually housing high-efficiency steam-electric units, which is normally operated to take all or part of the minimum load of a system, and which consequently produces electricity at an essentially constant rate and runs continuously. These units are operated to maximize system mechanical and thermal efficiency and minimize system operating costs (See Baseload).

Baseload Capacity: The capacity of generating equipment normally operated to serve loads on a round-the-clock basis (See Baseload, Baseload Plant), in megawatts electric (megawatts).

Basin (Sedimentary): A segment of the crust that has been downwarped. Sediments in the basin increase in thickness toward the center.

Brine: A highly saline solution. A solution containing appreciable amounts of sodium chloride and other salts.

Busbar Cost: The cost per kilowatthour to produce electricity, including the cost of capital, debt service, operation and maintenance, and fuel. The power plant "bus" or "busbar" is that point beyond the generator but prior to the voltage transformation point in the plant switchyard.

Capacity, Gross: The full-load continuous rating of a generator, prime mover, or other electric equipment under specified conditions as designated by the manufacturer. It is usually indicated on a nameplate attached to the equipment.

Capacity Factor: The ratio of the electricity generation of the generating unit, generating plant, or other electrical apparatus during a specified period of time, to the net capacity rating multiplied by the number of hours during the specified period of time (typically one year).

Capital Cost: The cost of wellfield development and plant construction and the equipment required for the generation of electricity from geothermal energy.

Centigrade: A common temperature scale in scientific work and throughout most of the world apart from the U.S. and U.K. To convert temperatures in degrees centigrade to temperatures in degrees fahrenheit, use the following formula: $F = 9C/5 + 32$.

Cogeneration: The sequential or simultaneous process in which steam is used to generate electricity and the associated heat is used in direct heat applications.

Combined Cycle: An electric generating technology in which electricity is produced from otherwise lost waste heat exiting from one or more gas (combustion) turbines. The exiting heat is routed to a conventional boiler or to a heat recovery steam generator for utilization by a steam turbine in the production of electricity. Such designs increase the efficiency of the electric generating unit.

Condensate: A heavier hydrocarbon occurring usually in gas reservoirs of great depth and high pressure. It is normally in the vapor phase but condenses as reservoir pressure is reduced by production of gas.

Conduction: Transmission through by means of a conductor. Distinguished, in the case of heat, from convection and radiation.

Connate Water: Water entrapped in the interstices of a rock at the time the rock was deposited.

Continental Drift: The concept that the continents can drift on the surface of the earth is because of the weakness of the suboceanic crust, much as we can drift through water.

Continental Crust: The type of crustal rocks underlying the continents and continental shelves.

Convection: Motion in a fluid or plastic material due to some parts being buoyant because of their higher temperature. Convection is a means of transferring heat through mass flow rather than through simple thermal conduction.

Convergent Plate Boundary: The boundary between two tectonic plates that are moving against each other.

Core: (1) The central region of the earth, having a radius of about 2,155 miles (3,470 km). Outside the core lie the mantle and the crust. The radius of the earth is 3,955 miles (6,370 km). (2) Cylinder of rock cut by a coring drill bit.

Crustal Zones: The outer layer of the earth originally considered to overlay a molten interior; now defined in various ways: lithosphere, sial, tectonosphere, etc.

Cycling: The practice of producing natural gas for the extraction of natural gas liquids and returning the dry residue to the producing reservoir to maintain reservoir pressure and increase the ultimate recovery of natural gas liquids. The reinjected gas is produced for disposition after cycling operations are completed.

Development Drilling: Drilling done in an ore deposit to determine more precisely size, grade, and configuration subsequent to the time the determination is made that the deposit can be commercially developed.

Downhole Measurements: Quantitative data gathered at various depths from man-made drill holes in the earth.

Earth Energy: Thermal energy at the normal temperature of the shallow ground (mean annual air temperature).

Enthalpy: A thermodynamic quantity expressed through the Second Law of Thermodynamics as the energy contained within a system.

Estimating: The process of forming an approximate judgement or opinion regarding the value, amount, size, weight, or other characteristic of a person, place, or thing.

Exploratory Drilling: Drilling to locate probable mineral deposits or to establish the native of geographical structure; such wells may not be capable of production even if mineral are discovered.

Exploratory Well: A well drilled to find and produce oil or gas in an unproved area, to find a new reservoir in a field previously found to be productive of oil or gas in another reservoir, or to extend the limit of a known oil or gas reservoir.

Fahrenheit: The common measurement of temperature in the United States. To convert temperatures in degrees fahrenheit to temperatures in degrees centigrade, use the following formula: $C = 5(F\text{-}32)/9$.

Fault: A plane of weakness within a rock body along which separation and differential movement occurs.

Fissure: An extensive crack, break, of fracture in the rocks.

Fluid Transport Medium: The liquid which transports energy, dissolved solids, or dissolved gases from their origin to their destination.

Fuel Cost: The monetary amount or the value of consideration-in-kind in goods and services given by the buyer to the seller to acquire fuel.

Fumarole: A vent from which steam or gases issue; a geyser or spring that emits gases.

Generation (Electricity): The process of producing electric energy from other forms of energy; also, the amount of electric energy produced.

Geology: The study of the planet earth including the materials of which it is made, the processes that act on these materials, the products formed, and the history of life since the origin of the earth.

Geophysics: Study of the features of the earth by quantitative physical methods.

Geopressured: A type of geothermal resource occurring in deep basins in which the fluid is under very high pressure.

Geothermal Energy: Heat energy from inside the earth which may be residual heat, friction heat, or a result of radioactive decay. The heat is found in rocks and fluids at various depths and may be extracted by drilling and/or pumping.

Geothermal Gradient: The change in temperature of the earth with depth, expressed either in degrees per unit depth, or in units of depth per degree.

Geothermal Plant: A plant in which the primary equipment is a turbine and generator. The turbine is driven either heat taken from hot water or by natural steam that derives its energy from heat found in rocks or fluids at various depths beneath the surface of the earth. The fluids are extracted by drilling and/or pumping.

Geothermal Recovery Rate: The rate of the proportion or percentage of geothermal energy extracted from the original geothermal reserves.

Geothermometry: The science of measuring temperatures below the surface of the earth.

Geyser: A special type of thermal spring that periodically ejects water with great force (See Thermal Spring).

Greenhouse Effect: A theoretical increased mean global surface temperature of the earth caused by gases in the atmosphere (including carbon dioxide, ozone, and chlorofluorocarbon). The greenhouse effect allows solar radiation to penetrate but absorbs the infrared radiation returning to space.

Groundwater: Water occurring in the subsurface zone where all spaces are filled with water under pressure greater than that of the atmosphere.

Heat: A form of energy related to the motion of molecules. Heat energy may be transferred from one body to another, as from the burner to the pan, to the water on a stove top. Heat energy may also be transformed into mechanical energy, for example, when heated gas pushes the piston in an internal combustion engine. Conversely, mechanical energy may be transformed into heat, for example, by pushing a piston to compress and heat a gas, as in a heat pump or air conditioner.

Heavy Metal: Metallic elements with high molecular weights, generally toxic in low concentrations to plant and animal life. Such metals are often residual in the environment and exhibit biological accumulation. Examples include mercury, chromium, cadmium, arsenic, and lead.

Hot Dry Rock: Heat energy residing in impermeable, crystalline rock. Hydraulic fracturing may be used to create permeability to enable circulation of water and removal of the heat.

Hot Spot: A localized melting region in the mantle a few hundred kilometers in diameter and persistent over long time periods. Its existence is assumed from volcanic or other geothermal activity at the surface.

Hydraulic Fracturing: Fracturing of rock at depth with fluid pressure. Hydraulic fracturing at depth may be accomplished by pumping water into a well at very high pressures. Under natural conditions, vapor pressure may rise high enough to cause fracturing in a process known as hydrothermal brecciation.

Hydroelectric Power: Electricity generated by an electric power plant whose turbines are driven by falling water.

Hydrothermal: Literally, hot water.

Identified Resource: Geothermal energy discovered through drilling, or by geologic, geochemical, and geophysical observation.

Igneous Rock: Rocks whose origin is the cooling and solidification of magma, molten rock material.

Injection Well: Well into which water or gas is pumped to promote secondary recovery of fluids or to maintain subsurface pressure.

Intrusion: A body of rock that has invaded the earth's crust from deeper depths in a molten state. Also, the process of this invasion.

Joule: A unit of work or energy in the metric system, equal to approximately 0.7375 foot-pounds.

Kaolinization: The chemical alteration process in which original rock elements are transformed into clay minerals.

Kilowatt (kW): One thousand watts of electricity (See Watt).

Kilowatthour (kWh): One thousand watthours.

Leadtime: Usually defined as the length of time from the start of construction of a facility to full-power operation.

Liquid-Dominated Geothermal System: A conceptual model of a hydrothermal system where hot liquids pervades the rock.

Levelized Cost: The present value of the total cost of building and operating a generating plant over its economic life, converted to equal annual payments. Costs can be levelized in current or constant dollars (i.e., adjusted to remove the impact of inflation).

Lithification: The process by which loose, unconsolidated earth material is converted to a coherent state.

Lithosphere: The upper, solid part of the earth. It includes the crust and uppermost mantle.

Load Following: Regulation of the power output of electric generators within a prescribed area in response to change in system frequency, tieline loading, or the relation of these to each other, so as to maintain the scheduled system frequency and/or the established interchange with other areas within the predetermined limits.

Magma: Naturally occurring molten rock, generated within the earth and capable of intrusion and extrusion, from which igneous rocks are thought to have been derived through solidification and related processes. It may or may not contain suspended solids (such as crystals and rock fragments) and/or gas phases.

Magmatic Intrusion: A body of magma that pushes its way into older rock.

Mantle: The layer of the earth lying between the crust and the core. The mantle extends between depths of about 19 miles (30 km) in the continental areas and 1,790 miles (2,800 km), where the core begins.

Marginal Cost: The change in cost associated with a unit change in quantity.

Market Penetration Model (Logistic Version): An exponential trend function in which the rate of growth begins at a low level, reaches a maximum, and then declines so that the increasing quantity approaches a maximum value (asymptote).

Megawatt (MW): One million watts of electricity (See Watt).

Methane: The most common gas formed in coal mines; a major component of natural gas.

Nonutility Generation: Generation by end-users, or small power producers under the Public Utility Regulatory Policies Act, to supply electric power for their own industrial, commercial, military operations, or for sales to a utility.

Operation and Maintenance (O&M) Cost: Cost of operating and preserving the physical condition and efficiency of plants used for the production, transmission, and distribution of energy.

Permeability: A measure of the capacity of a rock for transmitting fluid.

Plasticity: The property of a material that enables it to undergo permanent deformation without appreciable volume change or elastic rebound, and without rupture.

Plate Tectonics: A theory of global-scale dynamics involving the movement of many rigid plates of the earth's crust. Considerable tectonic activity occurs along the margins of the plates, where buckling and grinding occur as the plates are propelled by the forces of deep-seated mantle convection currents. This has resulted in continental drift and changes in the shape and size of oceanic basins and continents.

Plume: A body of magma that upwells in localized areas.

Porosity: The percentage of the volume of interstices or open space in a rock or soil compared to its total volume.

Public Utility Regulatory Policies Act of 1978 (PURPA): A part of the National Energy Act that contains measures designed to encourage the conservation of energy, more efficient use of resources, and equitable rates. Principal among these are suggested retail rate reforms and new incentives for production of electricity by cogenerators and users of renewable resources. The Federal Energy Regulatory Commission (FERC) has primary authority for implementing several key PURPA programs.

Quadrillion Btu: A standardized quantitative measure of energy in British thermal units (Btu) that allows comparison of different fuels and energy sources. The energy usage represented by this level is approximately 1 trillion cubic feet of natural gas or 170 million barrels of oil. Quadrillion is written with a 1 and 15 zeros.

Radioactive Decay: The spontaneous radioactive transformation of one nuclide to another.

Renewable Energy Source: An energy source that is regenerative or virtually inexhaustible. Typical examples are wind, geothermal, and water power.

Reserve: That portion of identified resources that can be produced legally at a cost competitive with other commercial energy sources.

Reservoir: A natural underground container of liquids, such as oil or water, and gases. In general, such reservoirs were formed by local deformation of strata, by faulting, by changes of porosity, and by intrusions. These, however, are classifications in the broadest sense.

Sea-Floor Spreading: The process by which molten mantle material rises at the mid-oceanic ridge solidifies and spreads the sea-floor laterally.

Seismic Activity (Seismicity): The likelihood of an area being subject to earthquakes. The phenomenon of earth movements; seismic activity.

Seismic Belt: An elongated earthquake zone.

Spreading Centers: The cracks in the lithosphere over extended distances above areas of upwelling.

Spreading Plate Boundaries: Zones associated with crustal plates on each side of a central rift zone, characterized by major rifts or faults and coincident with the world's mid-oceanic mountain and rift system.

Subduction Zones: An elongate region along which a crustal block descends relative to another crustal block. The depressing and passing of one plate margin of a tectonic plate of the earth under another plate.

Subsidence: Movement in the earth's crust in which surface material is displaced vertically downward with little or no horizontal component.

Tectonic Plates: Rigid lithospheric rock masses which form the uppermost portion of the crust.

Tectonic: Of, pertaining to, or designating the rock structure and external forms resulting from the deformation of the earth's crust. As applied to earthquakes, it is used to describe shocks not due to volcanic action or to collapse of caverns or landslides.

Thermal Energy: Heat energy.

Thermal Spring: Surface expression of groundwater in which the water temperature is at least 6.5°C greater than the mean air temperature.

Transform Fault: A strike-slip fault characteristic of mid-oceanic ridges and along which the ridges are offset. Analysis of transform faults is based on the concept of sea-floor spreading.

Transmission: The movement or transfer of electric energy over an interconnected group of lines and associated equipment between points of supply and points at which it is transformed for delivery to consumers, or is delivered to other electric systems. Transmission is considered to end when the energy is transformed for distribution to the consumer.

Transmission System (Electric): An interconnected group of electric transmission lines and associated equipment for moving or transferring electric energy in bulk between points of supply and points at which it is transformed for delivery over the distribution system lines to consumers, or is delivered to other electric systems.

Turbine: A machine for generating rotary mechanical power from the energy in a stream of fluid (such as water, steam, or hot gas), converting the kinetic energy of the fluid to mechanical energy.

Uncertainty: The state of not being definitely ascertainable or fixed as in time of occurrence, number, quality, or some other characteristic.

Undiscovered Resources: The presence of geothermal energy that has been estimated on the basis of geologic inference. Also termed unidentified resources.

Unidentified Resources: See Undiscovered Resources.

Vapor-Dominated Geothermal System: A conceptual model of a hydrothermal system where steam pervades the rock and is the pressure-controlling fluid phase.

Viscous Rock: Rock which flows in an imperfectly fluid manner upon application of unbalanced forces. The rock will change its form under the influence of a deforming force, but not instantly, as more perfect fluids appear to do.

Watt (Electric): The electrical unit of power. The rate of energy transfer equivalent to 1 ampere flowing under a pressure of 1 volt at unity power factor.

Watt (Thermal): A unit of power in the metric system, expressed in terms of energy per second, equal to the work done at a rate of 1 joule per second. See Joule.

Watthour (Wh): The electrical energy unit of measure equal to 1 watt of power supplied to, or taken from, an electric circuit steadily for 1 hour.

Wheeling: The use of the transmission facilities of one system to transmit power and energy by agreement of, and for, another system with a corresponding wheeling charge (e.g., the transmission of electricity for compensation over a system that is received from one party and delivered to another party).

Wheeling Service: The movement of electricity from one system to another over transmission facilities of intervening systems. Wheeling service contracts can be established between two or more systems.